How to Design and Install Outdoor Lighting

Writer
William H. W. Wilson

Designer
James Stockton

Photographers
Laurie Black
Michael McKinley

Ortho Books

Publisher
Robert L. Iacopi

Editorial Director
Min S. Yee

Managing Editors
Anne Coolman
Michael D. Smith

Senior Editor
Sally W. Smith

Production Manager
Laurie Sheldon

Horticulturists
Michael D. McKinley
Deni W. Stein

Associate Editors
Jim Beley
Diane Snow

Photo Editor
Pam Peirce

Production Editor
Alice Mace

Production Assistant
Darcie S. Furlan

Editorial Assistants
Laurie A. Black
Anne Dickson-Pederson
William F. Yusavage

National Sales Manager
Garry P. Wellman

Operations/Distribution
William T. Pletcher

Operations Assistant
Donna M. White

Administrative Assistant
Georgiann Wright

Address all inquiries to
Ortho Books
Box 5006
San Ramon, CA 94583-0906

First Printing in March, 1984

14 15 16 17 18
97 98 99 00

ISBN 0-89721-026-3

Library of Congress Catalog Card Number 83-62652

THE SOLARIS GROUP
2527 Camino Ramon
San Ramon, CA 94583-0906

Front Cover Photograph:
Uplighting emphasizes the dramatic impact of these redwoods.

Back Cover Photograph:
Top left: These Hydrangea plants are backlit to make them "glow".
Top right: A path light illuminates colorful flowers along this walkway.
Bottom left: White structures like this gazebo are major focal points in the nighttime garden.
Bottom right: A low path light highlights the texture of baby tears and blue fescue.

Title Page:
Outdoor lights create an inviting setting for evening entertaining.

Consultants:
Fred Brunswig. Brunswig Electric Co., Walnut Creek, CA
William Locklin. Loran, Inc., Redlands, CA
Harry White. Morris and White. San Antonio, TX

Acknowledgments:
Additional Photographs:
(Names of photographers and designers are followed by page numbers on which their work appears. R = right, L = left, T = top, B = bottom.)
Dennis Bettencourt: 32B; Richard Christman: Back Cover TR; Gerald L. French/PhotoFile: 46; Michael Landis: 34L, 66; Wayne LeNoue/PhotoFile: 47L; Lorraine Rorke: 12, 13, 64, 76; Tom Rosenthal: 63B; Carol Simowitz: 62–63T.

Illustrations:
Karen Tucker

Garden and Lighting Designs:
Chuck Archuletta, Rose Landscape Co.: 5, 19L, 35.
Artistic Lighting of San Rafael: 4 14, 18TL, 21B, 23TL, 23BR, 25, 52, 59L, 59R.
Thomas Baak & Assoc., Landscape Architect: 67
Donald G. Boos, Landscape Architect: 15, 18R, 24L, 71BL.
Boyd Lighting, Melanie Dane & Nancy Glen Designs: 36BR, 37BL, 37R, Back Cover TR.
Douglas Bungert, Landscape Architect; Front Cover.
Don J. Cantacessi, Landscape Architect: Back Cover BL.
Helen Reed Craddick, Interior/-Exterior Designer: 51, 88, 93R.
Dick Dubé. Landscape Lighting Studio: 58.
Robert Dye. Luxury Landscape: 38TL, Back Cover BR.
Fong & LaRocca Assoc.: 25B, 74.
Whisler-Patri, Architects and Interior Designers: 25B.
Jim Hagopian: 49, 54T, 55T.
Kotas/Pantaleoni, Architects: 24R.
Lifescapes, Landscape Architects: 23TR, 45.
Rudyard Morley. Distinctive Lighting: Front Cover, 17T, 21TR, 26, 28, 54B, 93L, Back Cover TL.
William J. Newton. Arbegast, Newton & Griffith, Landscape Architects: 20L, 21TL.
Royston. Hanamoto. Alley and Abey, Landscape Architects: Title Page, 57.
Art Thomas. AT Landscaping Inc.: 27.
William Wareham: 21TL
Harry A. White. Lighting Design: 8, 9, 10, 11, 16, 19R, 20R, 22, 55B, 69, 71T, 94.
Randall Whitehead. LIGHTSOURCE: 17B, 50, 68.

Color Separation:
Color Tech. Redwood City, CA

Typography:
Typothetae. Palo Alto, CA

Lighting Fixtures:

Cover	Loran, Footlighters
p.12	Nowell's, Anchor Light
p.14	Kim, BN-60
p.15	Jack Rydman, Oakville, CA, Model Napa 4 × 4
p.17B	Loran, Director
p.18TL	Kim, BN-60 (Path lights) Hubbell, 504 (Downlights and wall-washers)
p.19TL	American Lantern, Post Mount
p.21TR	Loran, Deliter
p.23BR	Kim, 309 (Path lights)
p.24R	Prescolite, AL-1
p.25T	Hydrel, 4436
p.25BR	Kim, PR43M (Underwater) Kim, C54 (Uplight)
p.32BR	Kim, "Wren House"
p.33TR	Kim, Well Light
p.33BL	Loran, Lawnlighter
p.33BR	Kim, Accent Light
p.35	Victor, V700-4
p.36TL	William B. McDonald, San Antonio, TX, Lattice
p.36TR	Loran, Illuminator
p.36BR	Stonco, Pagoda Light
p.37TL	Kim, Flower-Form Spread Light
p.37BL	Kim, Path or Border Light
p.37R	Malibu, Path Light
p.38TL	Loran, Niche-Liter
p.38TR	Kim, Wall and Sign Lighter
p.38BL	William B. McDonald, San Antonio TX, Model Tucson
p.50	Loran, Director
p.51B	Nova, C-82
p.53	Malibu, Halogen Floodlight
p.54B	Loran, Dolphin
p.55T	Victor, V103
p.58	Miniature Lighting, MH
p.67	Victor, V103
p.68TR	Loran, Director
p.71B	Victor, V103
p.74	Prescolite, 62030
p.75	Prescolite, 6203
p.82TL, TR, BL	Hubbell 587G
p.85BR	Loran, Maxilighter
p.86T	Loran, Illuminator
Back Cover, BR	Red Dot Lighting, G30

Special Thanks to:
Al Billings
Herb Burford
Nancy Christensen
Fillamento
Sandra Griswold
Heitmeyer/Wells Associates
Norman Johnson
Forrest Jones
Sue Fisher King
Steve Miller
Rud Morley
Roger Northrop
Nowell's Inc.
William Rogers
David Roman
Tuggey's Hardware
John Watson
Willie Wise
William Yusavage

How to Design and Install Outdoor Lighting

The Garden At Night

Lights bring your night garden to life

A garden in bright daylight is a play of color, texture, and form. Looking from the house, you can enjoy the rough grace of the gnarled tree arching over the lawn, the neat symmetry of green hedges, the many colors of brilliant flowers. The sun illuminates it all, hiding nothing.

At night, the same garden wears an air of mystery as light and shadow reveal some parts, hide others. The tree looms mysteriously above the grass. Moonlight captures the varied textures of a flowerbed. A wall or path is sensed in the darkness. There is magic and beauty in the night garden.

Well-designed garden lighting brings out the special beauty of the night garden. It illuminates the best features, leaving dark those you wish to conceal. Garden lighting shapes the view and mood; it can make a small garden seem spacious and attractive, a large estate intimate and comfortable.

Garden lighting also has a practical side. It helps you see where you're walking at night, and where any abrupt changes of level or hazards are. It discourages intruders by lighting the ground level clearly and eliminating shadows near the house. It permits you to use the garden after dark for relaxing, entertaining, playing games, or strolling.

You can design and build a garden lighting system yourself, using the methods discussed in this book. Learn about the techniques and materials of outdoor lighting in Chapter Two, the principles of lighting design in Chapter Three, and the process of installing and maintaining a lighting system in Chapter Four. Along the way, you'll find special sections on energy-efficient lighting, the use of colored lights, the best plants for lighting, outdoor insect control, and lighting for special occasions. Even if you plan to hire a lighting designer or other professional to do your outdoor lighting, the information in Chapter Two and the design process in Chapter Three will help you decide for yourself what you want to light and how it should be done, so that you can work with the designer or electrician in an informed way.

Garden lighting brings out a special beauty in the night garden (opposite), and expands your useable nighttime living area (right).

Lighting creates excitement and grandeur after dark.

LIGHTING AND THE SENSE OF SIGHT

Fully seven-eighths of the information we receive from our senses comes to us through the sense of sight. Light is essential to sight: we see things because light strikes them and is reflected. We are affected intellectually, emotionally, and physically by light, whether it is natural light from the sun or moon, or artificial light from lamps, light fixtures, and lighting systems. Though we can't control the light from the sun and moon, we can control what we see at night, using artificial lighting to selectively illuminate or conceal things, and to shape the way a place looks. Lighting is a powerful tool. It can show what is important in an environment, guide us to an entrance, and help us avoid hazards. Emotionally, lighting can make us feel happy and contented, or it can make us feel agitated, angry, or frightened. It can also affect us emotionally by illuminating the beautiful (or ugly) aspects of an environment. Physically, lighting can make us feel warm or cold, excited or relaxed, secure or uncomfortable. All of these effects are within our control when we design and build a lighting system.

LIGHTING TERMS

In planning and installing your outdoor lighting system, you'll need to know a few terms. Some of these are terms used to measure and describe light, others denote the parts of a lighting system. A few of the most commonly used terms are defined below; see the Glossary on pages 93–94 for others.

Absorption: A measure of the amount of light striking an object that is absorbed, rather than reflected. Surfaces that are black or dark-colored, and are heavily textured, absorb the most light.

Ballast: An electrical device that is used with fluorescent, mercury vapor, high-pressure sodium, and metal halide lamps to provide the power to start the lamp and to control the flow of electricity while it is operating. The ballast is usually built into the lighting fixture.

Brightness: Also called illumination, this is the amount of light striking a surface or object, measured in units called footcandles.

Energy efficiency: In lighting, energy efficiency is figured according to how much light (measured in lumens) is produced by one watt of electricity. See also pages 31 and 42.

Fluorescent lamp: A lamp with a coating on the inside that glows when activated by electrical current. Fluorescent lamps require special sockets and ballasts, and give an even, glare-free light.

Footcandle: The unit used to measure the brightness of the light striking a surface. Specifically, the amount of light that is cast on a surface one foot square from a standard candle a foot away.

Glare: Distractingly bright light, that interferes with our seeing what we need or want to see in an environment.

High-intensity discharge lamp: A lamp that produces light when electricity excites gases within a pressurized bulb. This group includes mercury vapor, metal halide, and high-pressure sodium lamps. All high-intensity discharge lamps require special fixtures and ballasts. Also called HID lamps.

Incandescent lamp: A lamp that produces light when electricity heats a metal filament. This is the lamp that most people think of as a "light bulb."

Lamp: The technical name for what we commonly call a light bulb. It is a tube, usually of glass, in which a filament, gas, or coating is excited by electricity to produce light.

Light output: The amount of light emitted by a lamp, measured in lumens.

Light source: The combination of a lamp and fixture that illuminates an environment.

Lighting fixture: The housing for a lamp, usually containing a reflector and electrical wiring connected to a power source. It may also contain a lens to control light spread and protect the lamp, and a ballast if it is intended for use with high-intensity discharge lamps.

High-intensity discharge lamps are most appropriate for large gardens and public places.

Lighting system: The complete system of lamps, fixtures, wiring, switches, and a power source, that supplies light to an environment. May also include auxiliary equipment such as transformers, time clocks, and sensors.

Low-voltage lighting system: A type of lighting that operates on 12-volt current rather than the standard 120 volts. (A few systems use 24-volt power.) Power is supplied by a transformer, which itself is connected to 120-volt power.

Lumen: A unit measuring the amount of light emitted by a light source. A lumen is the amount of light emitted by a standard candle.

Reflectance: A measure of the amount of light that strikes a surface and is reflected. Reflectance is highest from objects or surfaces that are smooth and light-colored.

Standard-voltage lighting system: The lighting fixtures and lamps that operate on standard 120-volt house current.

ABOUT LEVELS OF BRIGHTNESS

People differ in what they perceive as "bright" or "dim" light. Your own feeling about this will change according to the time of day, whether you're indoors or out, the relative brightness or darkness of the surroundings, and the type of lighting you are accustomed to.

Because of this, it is difficult to define a particular measured level of brightness (in footcandles) as "bright," "medium-bright," or "dim" light. But it is helpful to set some standards for each of these levels of brightness in approximate footcandles, so that you can better decide how bright to make each area of your garden and what types of lamps to use. The examples in each category below will help you to recognize the different levels of illumination in your everyday environment and determine which level is appropriate for each part of your garden.

For the purposes of this book, we will define "dim" light as light of .4 footcandle or less. Examples of dim light are full moonlight (about .1 footcandle), and the light on the sidewalk below a streetlight (about .3 footcandle).

"Medium-bright" light is light with a brightness of .5 to 8 footcandles. This level of illumination is brighter than moonlight or the light of streetlights, but less bright than the lighting inside your house. The amount of illumination given by low-voltage garden lighting fixtures is near the lower end of this range, and most standard-voltage garden fixtures illuminate to a brightness of 3 to 5 footcandles.

We will define "bright" light as light of 8 footcandles or greater. Many indoor environments are lit to this level of illumination. For instance, a small reading lamp illuminates a book to about 30 footcandles, and the level of lighting in most offices is between 70 and 150 footcandles.

It is rarely necessary or desirable to illuminate the home garden to a greater brightness than 20 footcandles; in fact, 5 footcandles may be a good maximum level for most gardens. When garden lighting is too bright, problems can arise with glare unless the fixtures are extremely well shielded, and the brilliant light may bother neighbors.

Changes in the level of brightness from daytime (below) to nighttime (right) can create dramatically different moods in an outdoor setting.

The level of brightness should be varied. Use brighter lighting along walkways for safety and dimmer lighting in more intimate areas.

OUTDOOR LIGHTING THROUGHOUT HISTORY

Lighting the outdoors began when our remote ancestors first tamed fire for light and warmth. Torches made of resinous wood lit the path at night, and streamside rushes dipped in animal tallow made simple candles. These "rushlights" were succeeded eventually by wax candles. The Emperor Constantine, in the fourth century, is said to have lighted the streets of Constantinople with thousands of candles, until it was as bright as day.

Since ancient times we have used lanterns to shield oil lamps and candles outdoors. They have been made of wood, stone and metal, and carved or pierced to emit light. Later lanterns had "windows" of horn or mica or—eventually—of glass. The Japanese have used stone and metal lanterns for centuries to illuminate gardens, and to light the entrances to shrines. Lanterns containing oil or tallow were still used in Europe in recent centuries to light homes, carriages, and ships.

The most important modern developments in lighting have occurred since the beginning of the last century. In about 1800, a process was developed to recover a gas from the distillation of coal. This product revolutionized lighting, and gas soon lit the streets of London and other cities. The heyday of gaslight lasted most of the century; it ended with the development of electric lighting.

The first type of electric light developed, electric arc lighting, still finds specialized applications today. It was used for streetlighting in some areas, but was soon succeeded by the incandescent light bulb, which was invented by Thomas Edison and others in about 1880.

Electric lighting has made possible garden lighting as we practice it today. Garden lighting is really a recent development, largely taking shape since the invention of the incandescent lamp. Many developments important to outdoor lighting have followed since that first electric bulb, among them incandescent lamps for use outdoors, mercury vapor and other high-intensity discharge lamps, and low-voltage lighting systems; new products for outdoor lighting continue to be introduced. Many developments have also been made recently in lighting theory as researchers and designers study the effects of lighting on people and environments and evolve an increasingly sensitive approach to lighting design.

An old ship's lantern brightens a modern patio during the day and night.

MODERN PERSPECTIVES ON LIGHTING

The efforts of professionals in many fields have brought about these developments in outdoor lighting. Landscape designers, architects, and even theatrical lighting experts have helped develop new lighting materials and design theories, but the strongest influences have been those of illuminating engineers and lighting designers. The expertise and perspectives of these two fields differ, yet each has been important in developing outdoor lighting.

Illuminating engineers measure the amount of light emitted by a lamp and received by a surface. They design lighting systems to provide a certain quantity of light with the greatest energy efficiency. The lighting designer, on the other hand, considers how lighting shapes and reveals the features of an environment and how it affects the people who use that space. Designers work not only with the intensity of light needed for particular tasks, but also with the patterns of lighting. They use light to indicate what is important in an environment and even to affect the emotional states of its users.

A Balanced View

This book is based on the insights offered by illuminating engineers and lighting designers, as well as other lighting professionals. The lighting materials discussed in Chapter Two have been developed mainly by lighting engineers, and we've tried to include some of the most recent innovations, such as the new, smaller HID lamps, the powerful low-voltage spotlights and floodlights now available, and alternative lamps and fixtures developed for energy efficiency.

The lighting techniques in Chapter Two and the design process in Chapter Three are based on the ideas of lighting designers, who have generously shared them.

Getting Help

You can use this book to design and install a complete outdoor lighting system yourself. If, however, you do not have the time or energy to do the work, you can choose instead to work with a lighting designer or electrician. These professionals can be especially helpful in designing and installing complex systems with sophisticated controls, but they can assist on smaller projects as well.

Even if you will be hiring others to do the work for you, you should first learn all that you can about outdoor lighting. This is important because many lighting designers and electricians are trained to work primarily with indoor lighting, and may lack experience in lighting gardens. Read Chapters Two and Three of this book to determine what you want to light in your garden, the relative light intensities and different lighting techniques you want in different areas, and the types of lamps and fixtures you prefer. You will then be able to work with the designer or electrician to get the lighting system you want.

Working with a Lighting Designer

Lighting designers can be very valuable on large, involved lighting projects. A designer can provide you with information on the local codes, proper electrical loads, available lighting fixtures and hardware, low-voltage lighting, and high-intensity discharge lamps. A designer can create a unified design that is sensitive to your needs and enhances your garden.

Check the credentials of any lighting designer you're considering working with. Many designers are most experienced with indoor lighting but may have done successful outdoor projects as well. Find out about the designer's experience, ideas about the place of lighting in outdoor design, and other projects the designer has done that you can go see. Try to sense whether the designer's approaches will work for you.

Designers are usually paid as hourly consultants and should inform you of their rate schedule when you first meet.

If your lighting is a part of a larger home or garden development project, involve the designer in the planning process as early as possible. Discuss the design concepts for both indoor and outdoor areas, your living habits, and your ideas and needs regarding outdoor lighting.

Electrical Contractors

If your problem is technically complicated, an electrician can be essential. For instance, a system that includes remote-control relays or extensive wiring around water may be beyond your capability. On smaller projects, too, the electrician can provide valuable expertise at difficult moments in the installation to help you get it done. If you are not comfortable doing 120-volt wiring, the electrician can do all of the wiring for you, or perhaps just such tasks as installing switches and controls, wiring lighting for a hot tub, or connecting the new system to the power source.

An electrician is an expert on safe wiring practices for all types of installations, and probably knows the electrical code as well as your local inspector. Even if you do all your own work, we recommend that you have an electrician check it at two points in the process. First, have your final lighting plan and the specifications for fixtures, lamps, and wire or cable checked, especially for safe electrical loads and wire runs. And have the electrician check the completed system as you finish your installation work to be sure all wiring and connections are done safely. This second check by the electrician may not be necessary if the system will have a final inspection by your local building inspector.

Plan to pay the electrical contractor the hourly labor rate for any installation work done and an hourly consultation rate for checking your plans and the finished outdoor lighting system.

If you live in a large urban area, there may be electricians and lighting designers available who specialize in outdoor lighting. These professionals may be the most qualified to help you with your outdoor lighting project, but you should still ask to see other jobs they have done to decide if their particular style matches your needs.

A lighting designer can help you create a unified design.

Tools and Techniques

Create a variety of effects by choosing the right techniques

The prominent architect and lighting designer Richard Kelly observed that people see lighting in a very simple way. We react, he said, to only three kinds of light, which cause all the different visual effects we perceive around us. He called these three kinds of light Focal Glow, Ambient Luminescence, and Play of Brilliants.

Focal glow, he wrote, is light that commands attention, fixes the gaze, and creates interest. It is like the light burning at the window, or the shaft of sunlight that warms the far end of the valley. It tells us what is important, directs and controls our attention, and helps us see.

Ambient luminescence is the illuminated background. It creates the comfortable room in which focal glow shines. It is like a snowy day in open country, or being underwater in the sunshine. It has the effect of minimizing everything and unifying all. It is restful.

But we need also the sparkle of a tumbling brook, the heaven full of twinkling stars, and the flicker of candle flame. This active, exciting light is the *play of brilliants.* It stimulates us, charms the senses, and makes us feel alive.

You can create all these kinds of light in your garden with the techniques and materials discussed in this chapter. Each technique can bring alive particular beauties of your garden. The materials you can use include 12-volt and 120-volt systems, different types of lamps and fixtures, and the other materials that connect these parts into a whole.

Opposite: Garden elements needn't be large to be interesting at night. Placing a fixture close to the ground illuminates small features such as rocks, ground covers, and other small plantings.
Right: Background lighting provides the ambient luminescence and a bollard light adds focal glow to this outdoor "room."

LIGHTING TECHNIQUES

Many techniques are used in lighting gardens. They vary in the position of the light source and whether it is aimed up or down, or across a surface. Different techniques can bring to your garden the spotlight of focal glow, a background of ambient luminescence, or the sparkle of a play of brilliants. Choose from these techniques to light the different parts of your garden, and combine them to make your outdoor lighting system.

Downlighting

This is a general term and includes many of the other techniques given below. It is the lighting of an object, area, or surface from above. The light source can be a large floodlight to provide general illumination for safety, security, or entertainment (see Security Lighting, Safety Lighting, Area Lighting), it can be several smaller floodlights set high in a tree, giving filtered light (see Moonlighting), or it can be diffused through a globe, stretched canvas, or other diffusing material (see Diffused Lighting). It can also focus on a particular plant or small area in the landscape (see Spotlighting, Accent Lighting, Contour Lighting), or help to suggest background, foreground, and perspective in a garden (see Perspective Lighting).

When done well, downlighting imitates nature. It can suggest the morning sun gently lighting the patio, a shaft of sunlight piercing the clouds on an overcast day, or full moonlight filtering through the trees.

Below: Downlighting combined with a gentle breeze adds life to the shadows in this garden.

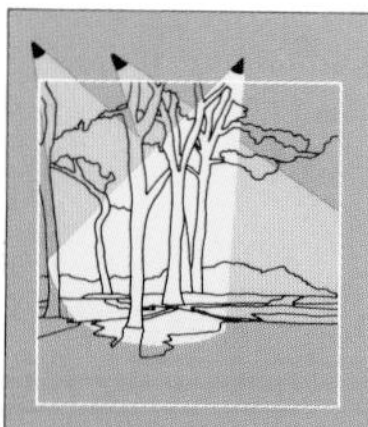

Left: Bold furrowed bark and drooping branches are good candidates for uplighting.

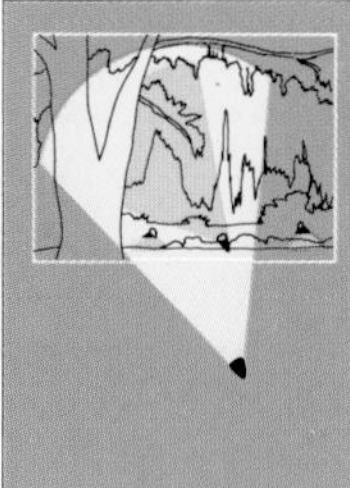

Uplighting

Also a general term, this means lighting something from below. Shadowing, mirror lighting, and silhouetting are all types of uplighting. Uplighting provides focal glow in the garden. It is dramatic lighting, akin to the spotlight in a theater or searchlights crossing the sky at night. Uplighting is rarely seen in nature, and this gives it a slightly unnatural appearance, demanding attention. Use this technique with discretion—when there is a reason to call special attention to a part of the garden or to a particular feature within it, such as a statue, a large and striking tree, or a house facade or wall. Uplighting can be seasonally spectacular too, illuminating autumn-colored branches or the swirling snow of a snowstorm.

Safety Lighting

One of the most important jobs of garden lighting is to illuminate garden spaces to give directions to its users about obstacles to avoid and where to walk. In other words, it allows people to feel comfortable and move easily in the garden. This is the purpose of safety lighting. You can light for safety with well-planned downlighting from trees or nearby buildings (see Moonlighting), or with low path lights, such as the common "pagoda" or "mushroom" shaped fixtures. Safety lighting should be brightest on heavily used paths or steps. It also needs to be brighter if areas nearby are brightly lit or if it will seem dim by comparison to other lighting.

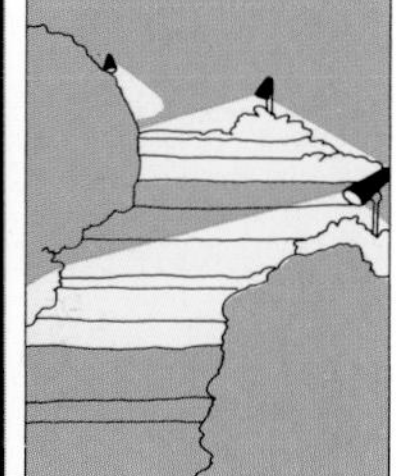

Left: Low path lights provide safety lighting without sacrificing the attractiveness of this rustic stone entryway.

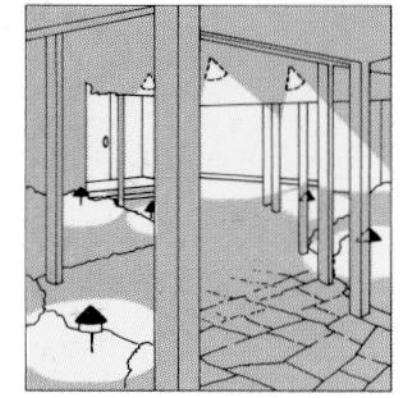

Effective security lighting illuminates areas evenly, especially entryways, without leaving pockets of shadow.

Security Lighting

Many people consider protecting their property from burglars and other marauders a major goal of garden lighting. In fact, homeowners often choose to put in security lighting before any other outdoor lighting.

Much harsh and unpleasant lighting design is done in the name of security lighting. Lights installed on the house eaves and aimed into the eyes of visitors are an example of this. Proper illumination of the paths, plants, and areas of the garden protects your home from unwelcome guests while enhancing its beauty. Security lighting does not need to be glaring and vicious to be effective.

An important ingredient in security lighting is effective switching. The switches should be placed indoors in a central location, and should allow you to operate different parts of the system separately. Many devices are available for this, including plug-in indoor switches, time clocks and photocells to automatically turn the lights on, and wireless controls. See page 42.

Area Lighting

Good area lighting allows you to entertain on the deck or patio, play games on the lawn, or even do light gardening work after dark. Area lighting is ambient luminescence in the garden that provides the background for focal and accent lighting. It is usually done by downlighting from overhead with floodlights or diffused low-voltage lamps. When using floodlights in the garden, whether for area lighting or security or safety lighting, keep in mind that floodlights alone give a flat, dull look to the garden in which no special features stand out. And if the floodlighting is too bright, it can ruin any subtle lighting effects nearby. To avoid this, combine floodlighting with more decorative types of light. You can filter the brilliance of the lamps through trees, as in moonlighting. You can also use spotlighting or accent lighting nearby to call attention to special areas and allow other areas to be dim or dark, providing contrast. Floodlights near sightlines in the garden can also be major sources of glare, which is distracting and disturbing to the eye. As much as possible, the lamp and fixture itself should be placed high above ground, or at least partly concealed from view by filtering or diffusion of the light in some way.

Area lighting for outdoor night games needs to be bright and uniform, but it shouldn't glare into your eyes and it shouldn't shine into neighbors' yards.

For outdoor living, use even, general illumination on decks or patios.

Diffused Lighting

You can diffuse area lighting by positioning the light source behind some translucent object such as frosted glass, a canvas panel, or simply the globe or plastic panel supplied as part of the fixture. This gives the light a glare-free quality with softer shadows.

Diffused lighting is useful for outdoor dining areas, quiet areas of patios and decks, and other areas used for relaxation and passive recreation.

Moonlighting

Moonlighting uses a mild source of light, usually positioned high above the ground, to simulate the soft, diffuse light of moonlight. A number of prominent lighting designers work primarily with moonlight effects, using this soft, attractive light to create focal points, give soft background luminescence, cast attractive patterns of light and shadow from large trees, and provide for security and safety in the garden. They feel that this artificially created moonlight is the most natural-looking outdoor light.

The higher above the ground the light source for moonlighting is, the better. It is important that you don't see the light source itself, but only its effect as it selectively illuminates tree branch and trunk and casts graceful shadows on the ground.

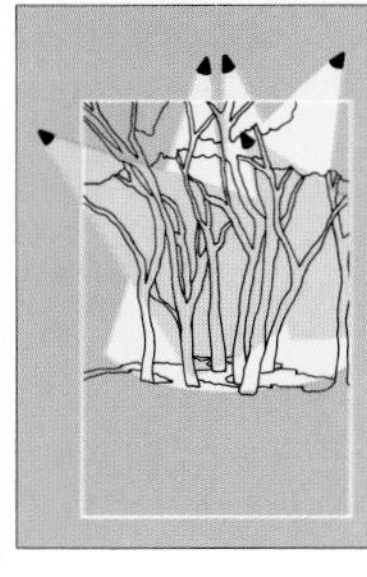

Above left: Diffusing the light from an eye-level fixture avoids glare. Above right: Light fixtures placed high in trees create the effect of moonlight while providing light for activities, safety, and security.

Grazing light emphasizes the texture of a wall or paving.

Grazing Light

A smaller-scale effect, grazing light is the positioning of a light source to bring out the texture of an interesting garden surface, such as a masonry wall, an interesting paving, an attractive door or fence. Place the light source several inches from the object and align it to direct light across the surface. Uplighting and downlighting of trees can use grazing effects to bring out the texture of the bark.

Cross Lighting

When an area or object is lit from two or more points by cross lighting, it is revealed more fully than when it is lit from only one direction, and the shadows are softer. This technique is used in downlighting or moonlighting in trees, illuminating statues, or spotlighting specimen plants. Cross lighting is usually done with broad floodlights or other diffuse sources of light, rather than with spotlights. When used in downlighting, cross lighting works best if the beams of light cross high overhead, rather than near the ground.

Cross lighting creates attractive shadows in this "moonlit" garden.

Left: Spotlighting makes this unique statue an important focal point in the night garden.
Below: Low-voltage lights are ideal for accenting individual plants.

Spotlighting
Spotlighting gives focal glow to a garden, directing an intense beam of light to pick out a particular garden detail or object. Spotlighting of important focal points in a garden, done discreetly, can be very effective. It reveals what is important in the garden. If overdone, it turns the garden into a circus. Some frequent uses of spotlighting in gardens are in tree uplighting, lighting statues, or dramatic illumination of a house entry. In combination with floodlights it creates interesting patio lighting or illuminates special small areas in the garden.

Accent Lighting
This term is used to include the many small-scale lighting techniques that add focal glow and sparkle to garden lighting. One type of accent lighting uses small spotlights to bring out particular plants, other small garden elements, or works of art. Other types of accent lighting give sparkle to the garden with twinkling lamps or low decorative fixtures placed at the focal points. Accent lighting frequently uses low-voltage fixtures and lamps, and is often more appropriate to a small garden than brilliant spotlighting or floodlighting.

Silhouetting
Silhouetting an object against a lighted wall creates a dramatic effect. You can see this type of lighting in nature in the silhouettes of trees along a ridge at dusk. In garden lighting, the shadow of a distinctive plant, stone, or other object is seen against a wall or other vertical surface that is lit from below. The light source should be concealed, perhaps sunk in the ground or placed directly behind the object being silhouetted. Plants with distinctively shaped leaves, or which move freely in a wind or storm, can be especially attractive lit in this way.

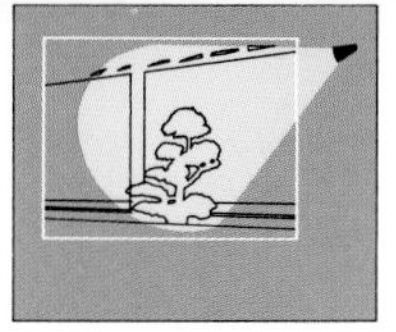

Left: Plants with distinctive shapes are especially striking when seen in silhouette.

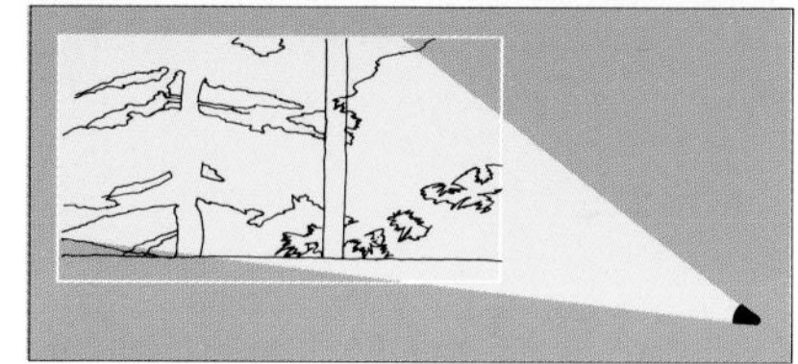

Shadowing
Shadowing is similar to silhouetting, except that the shadow of the plant is projected on the wall or other vertical surface by being lit from the front. Usually the source of light is low to the ground, in front of the plant, stone, or other object, aiming up through it toward the wall.

Contour Lighting
Contour lighting helps orient the visitor in the garden and provides for safety. It usually consists of downlighting, often created with narrow floodlights or low area lights, that illuminates the meeting place of lawn and ground cover, for instance, or shrubs and the surface of a garden pool, paved driveway and lawn, and so on.

Fill Lighting
All lighting schemes need to provide contrast by combining bright focal glow with other areas of dim ambient luminescence, or even with darkness. Fill lighting provides the dimly-lit backdrop for the brighter garden focal points. Ideally it is accomplished as excess light from these focal points spills out into the surrounding area. Low-wattage area lights such as path lights can provide fill light by dimly lighting the floor of the garden. To stand out strongly, the lighting in focal areas should be about ten times as bright as the surrounding fill light.

These small, low-voltage lamps add sparkle to the landscape.

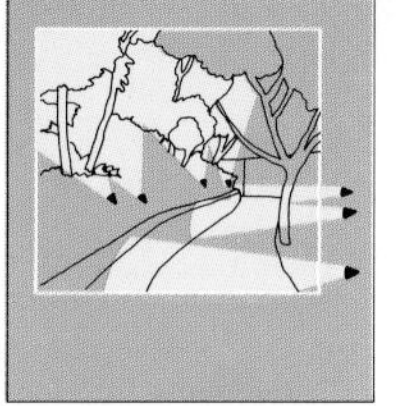

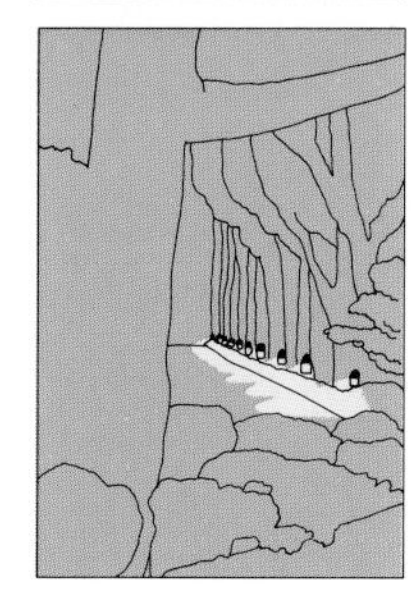

Top, left: Shadowing is even more dramatic than silhouetting, since the shadow of the object is often greatly magnified.
Top, right: Light spilling from the spotlights illuminates this path.
Right: Contour lighting defines the driveway and adds depth and perspective to the garden at night.

Background Lighting
Background lighting provides a visual backdrop for the garden, or a small part of it. You can do this with perspective lighting, or you can light a stone wall, other structure, or tall, dense planting at the back of the garden. Both background lighting and perspective lighting provide effective vistas for viewing from indoors.

Vista Lighting
Many gardens, especially those in hilly areas, look out on a beautiful view. This may be a natural area, such as a woodland, field, or nearby mountain, or it may be the myriad lights of a city. Vista lighting is a way of protecting that view, of using it to form the garden backdrop by night as well as by day. This is done by carefully controlling the height and brightness of garden lights. For example, you may use medium-bright focal and area lighting in the garden foreground near the house, but only low-voltage area lights in fixtures less than two feet high in the farther garden, so you can see the view beyond. Or if tall trees frame the view, uplight them from the ground to emphasize the view.

Above: These background trees form a glowing "wall" that helps light this deck. Right: The lights on this deck are directed downwards so they won't disrupt the vista.

Perspective Lighting

Perspective lighting gives a subtle sense of drama and dimension to the garden by emphasizing a line of sight (called an axis). A garden axis may be a natural view through a group of trees, the open corridor through plantings formed by a path or a long, narrow patio. Frequently, perspective lighting is used to make a garden feel larger than it actually is by suggesting to the viewer's eye that the other end of the visual axis is at some distance. One way to do this is to light the foreground (from which you view the garden) dimly, provide only dim lighting along the visual axis, and light the focal point at the end of the view brightly.

Water Lighting for Garden Pools, Fountains, and Streams

Small pools and other garden water elements can be attractive when lit from underwater, but two cautions apply here. First, do not plan to light from underwater if you aren't sure you can keep the water reasonably clear. Murky water looks murkier when lit, because all the small particles of suspended matter catch and reflect the light. Second, it is important here, as in all outdoor lighting, to conceal the light source. The section on water lighting in Chapter Three offers suggestions for doing this (see page 54).

It is easiest to install the wiring and lighting fixtures for water lighting when building the pool. If you're lighting an existing pool, you can get waterproof portable fixtures on cords.

Top: The rows of lit Italian cypress trees add depth to this garden by emphasizing the distance to the cabana.

Right: Underwater lighting and a grazing uplight create stark drama at this fountain.

Water Lighting for Swimming Pools

Position swimming pool lights on the side or end from which the pool is most often viewed, so they will not glare in your eyes. Install dimmers on this circuit so you can use the pool lights as soft uplighting or as an accent light when you're not actively using the pool or turn the lights up for exciting swimming.

Mirror Lighting

Water in the garden can also be used as a mirror. A garden pool or swimming pool can beautifully reflect buildings and trees. To do this, light an area that lies behind the reflecting surface, in the line of view, while leaving the water itself dark. Large specimen trees near a swimming pool may be uplit with floodlights or spotlights. Subjects may include specimen trees and shrubs, Japanese lanterns (lit with a small incandescent lamp or candle inside), or votive candles around a small pool (see Special Occasion Lighting, page 60).

Above: Mirror lighting can be spectacular, especially when large trees like these redwoods are being reflected.
Right: Swimming pools are usually lit from underwater with large fixtures and bright lamps built into the pool when its shell is formed.

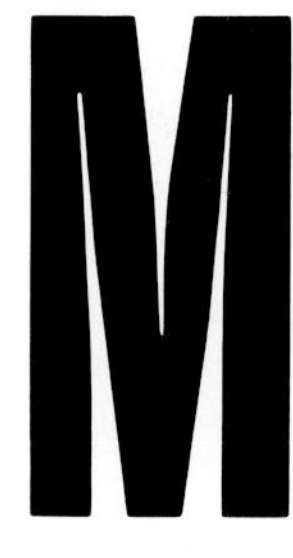

ATERIALS FOR LIGHTING

A wide array of lighting fixtures, lamps, and other materials has been developed by lighting engineers and designers. With these materials, you can bring any lighting technique described above into your garden, light the plants and materials brilliantly or softly, and control the lighting system easily, even automatically.

One important recent development in lighting materials is the low-voltage lighting system. These systems usually run on 12 volts of power rather than the standard 120 volts supplied to your home. A number of these systems are available to homeowners and can be very economical to install. If well used, low-voltage lighting can provide beautiful, useful lighting.

You can install an outdoor lighting system within almost any budget. The least expensive low-voltage lighting systems should cost you less than $150, and you can install a simple standard-voltage system for a few hundred dollars. Of course, you can also spend thousands of dollars on a large and elegant outdoor lighting system, especially if it includes high-intensity discharge lamps, top-quality fixtures, and refinements such as remote-control switching.

Refer to the chart on 12-volt and 120-volt systems to help you decide whether to use low-voltage lighting in your garden, or to determine the extent to which you'll use it. Each type of system has its distinct advantages. Which one you'll use depends on the size of your garden, your budget, and your skill in electrical wiring. And you do not need to be exclusive; low-voltage and standard-voltage systems can work very well when combined in a garden, with each serving its particular purpose.

The soft light from a 12-volt lighting system keeps this small pool setting intimate.

12-VOLT AND 120-VOLT LIGHTING SYSTEMS

1. Safety, ease of installation
Low-voltage systems are usually very safe to install; the only potential dangers are in the connection to the 120-volt power (if you can't simply plug in the transformer), in using 12-volt wire of too small a size, or in making improper 12-volt connections.

Standard-voltage systems must be connected to the power source carefully (see page 78). It is important to calculate electrical loads properly, follow the electrical code, and have the system inspected before connection. Hire an electrician for part or all of the work, if you are not completely comfortable with it.

2. Levels of brightness
Low-voltage levels of brightness range from dim to medium-bright (from less than .5 footcandle to 8 footcandles). However, low-voltage systems can be especially bright if you use the new quartz incandescent 12-volt lamps. The area lighted to medium-bright intensity by each lamp is smaller than for most standard-voltage lamps. You need more lamps to achieve the same level of brightness as with standard-voltage systems.

Standard-voltage levels of brightness range from dim, with small lamps and dimmers, to brilliant (too bright for many purposes) with some quartz, mercury vapor, and metal halide lamps. Some lamps will light thousands of square feet.

3. Cost to install (if you're doing it yourself)
Low voltage is relatively inexpensive. A set including a combination transformer and time clock, 100 feet of wire, and six plastic floodlights may cost less than $150.

Standard-voltage systems can be relatively inexpensive to very expensive, depending on the quality of fixtures and the size of the system. For 100 feet of Type UF wire, six fixtures, and two switches, a standard-voltage system might cost about twice as much as an inexpensive low-voltage system, but the materials may be more durable. A standard-voltage lighting system may also cost you more because you may hire an electrician to install part of it.

4. Maintenance needs and expense
Low-voltage systems require annual lubrication of lamp sockets, periodic checking of fixture position, especially for spike-mounted fixtures, cleaning out of debris from uplights, and relocation of fixtures or tree pruning. Electrical problems may be less serious because of the lower voltage.

Standard-voltage systems require annual lubrication of the lamp sockets, removal of debris, pruning around ground-mounted fixtures, and periodic checking of cables in trees. Fixtures may need to be relocated, especially after tree pruning or severe storms.

5. Durability of components
Durability of fixtures, wire, and other components on both types of systems may be poor to excellent, depending on the quality of the materials, performance of routine maintenance, and the demands of the environment. (See page 87.) Some low- and standard-voltage fixtures are made of thin, brittle plastic, while others are of aluminum alloy, enameled to resist weather, and lasting far longer. Some expensive fixtures made of brass and heavy cast aluminum may last the longest. Spend as much as your budget allows for the fixtures and materials; they should pay you back with long life.

6. Ease of fixture relocation
Low-voltage systems are easier to relocate. The cable is shallowly buried, if at all, and conduit is rarely used. This also means that fixtures may occasionally relocate themselves by falling over or tilting.

Standard-voltage systems are more difficult to relocate. Relocation requires cutting the conduit or at least removing the concrete base at the fixture, digging up deeply buried cable, and retrenching. Relocation is easier with tree fixtures and with those mounted on portable stake mounts.

7. Special requirements
Low-voltage system requirements for connection to switches and power source are identical to standard-voltage systems, but low-voltage systems also require a transformer and 12-volt lamps.

Standard-voltage systems require electrical conduit, at least above ground. Check your local code for other requirements.

8. Best uses
Low-voltage systems are best in smaller gardens because of the smaller area lighted by each lamp. The new low-voltage quartz lamps make larger and brighter spotlight and floodlight coverage possible. Low-voltage lighting works well on smaller garden subjects, such as medium and small trees, shrubs, ground cover, small statues, and area lighting for small garden areas.

Standard-voltage systems are more valuable in areas where brilliant illumination is required for safety or security (most of these are public areas; residences can do this with less light). They are also good in spots where you must light from farther away, as in uplighting huge trees.

In sum, your choice of low-voltage or standard-voltage lighting systems mainly depends on the size of the area or object to be lighted, the level of brilliance you want, and your budget. A large landscape may be more efficiently lit by fewer, standard-voltage lamps, but a small townhouse or apartment garden may be made impossibly bright by 120-volt lamps, so it is a good site for 12-volt lighting. Also, if you are inexperienced with electricity and wish to install your own lighting system, particularly if you are on a low budget, low-voltage lighting may be your best bet.

Lamps

Technically, a lamp is what you've always called a "light bulb": a tube, usually of glass, in which a filament, gas, or coating is excited by electricity to produce light. When choosing lamps for garden lighting, you have many choices. The type of lamp most used in the past has been the incandescent lamp, from small lamps used in accent and path lights, to bright lamps of several hundred watts for lighting large outdoor areas and underwater lighting in swimming pools. Several other types of lamps are available, including low-voltage and quartz incandescent lamps, fluorescent tubes, and various high-intensity discharge lamps, among which are mercury vapor, high-pressure sodium, and metal halide. Of these, both mercury vapor and metal halide lamps have some uses in lighting home gardens. High-pressure sodium is highly energy-efficient, but is less frequently used for residential locations because of its orange color. New sodium lamps are currently being developed in lower wattages and with better color rendering.

When choosing lamps for your garden, plan to use primarily one type, whether low-voltage incandescent, standard incandescent, or mercury vapor. Fluorescent tubes and metal halide lamps, if used, are best in accenting primary areas and focal points, rather than for general illumination of the garden. This is because fluorescent light used alone in a garden is flat and without sparkle, while metal halide lamps are designed to give brilliant light for large areas. Use brighter lamps where you need them to provide outdoor focus for active night use or for safety; use lower wattage lamps for soft area lighting, moonlighting, and fill-in lighting.

When selecting the best type of lamp for your situation, you must consider many factors. Durability, cost, and energy efficiency are most important in designing large systems, because the expense of running the

Clockwise from left, an ellipsoidal reflector lamp, an incandescent lamp, a reflector flood lamp, a PAR lamp, a high-pressure sodium lamp, a low-voltage quartz halogen flood lamp, four low-voltage lamps, a low-voltage flood lamp, and tungsten-halogen quartz lamp.

system, and of replacing burned-out lamps (especially if they're located in difficult spots) can be considerable. A more important consideration in the small garden is the range of sizes available; incandescent lamps come in a tremendous range of sizes to fit every need, but metal halide lamps do not. Color rendering is also important, because it has a lot to do with how the lighting makes the garden look and how it makes you feel.

Lamp Types

Here are the major types of lamps used for outdoor lighting.

Incandescent lamps give light from a glowing tungsten thread or filament within the glass bulb. They include the standard "A" type bulb used commonly indoors; reflector bulbs, designated as R (for reflector), ER (for elliptical reflector), and PAR (for parabolic aluminized reflector lamp, the most commonly used bulb for outdoor spotlighting and floodlighting). Also included are low-voltage versions of A, R, and PAR bulbs.

Quartz incandescent lamps work on the same principle as standard incandescent lamps, but operate at a higher temperature and give more consistently bright light.

Fluorescent tubes are not frequently used in garden lighting, but can be valuable for certain uses, such as lighting of signs, bannisters, and uplighting of walls.

Mercury vapor, high-pressure sodium, and metal halide lamps belong to the family of lamps known as high-intensity discharge lamps. When turned on, the electrical current excites specific gases inside these pressurized bulbs. The bulb may then take several minutes to warm up before it gives full light. Both efficiency and color rendering are sometimes factors in deciding whether or not to use high-intensity discharge lamps. Note that although several types of mercury vapor lamps are available, clear mercury lamps (also called "blue") are most common in garden lighting.

Lamp Characteristics

Here are some of the most important characteristics of these lamps.

Efficiency is the approximate amount of light output, measured in lumens, per watt of electrical input. (See also the notes on energy efficiency, page 42.)

Efficiency is most important to consider with large systems. With small lighting systems, the amount of energy and money that can be saved by substituting more efficient types of lamps may not outweigh negative factors like poor color rendering. But increasingly, manufacturers are developing more energy-efficient lamps of all types, and you can try to include these in your system, no matter what type of lamp you plan to use. For example, 65- and 120-watt incandescent floodlights and spotlights are now available that put out about as much light as the old 75- and 150-watt lamps, while using less power. And a new 95-watt quartz lamp with a PAR lamp shape emits as much light as a 150-watt PAR incandescent lamp on less energy. See also the notes on energy-efficient lighting systems on page 32.

Lamp sizes and light output determine how bright your garden lighting will be. Lamps are usually sold according to the number of watts of electricity they consume. Light output is measured at the lamp. The lumen output of most lamps diminishes with age—that is, they give less light while using the same amount of electricity, so that after a while it may be best to replace an old lamp. The values given are for the initial lumen output of new lamps.

The type and sizes of lamps you choose for your lighting system, along with how you position them and the types of fixtures you choose, will determine how bright your lighting system will be. How bright you want it depends on what you want to light, how you use the garden at night, and your personal taste. You may find that you enjoy the soft lighting provided by low-voltage systems or small standard-voltage lamps, or you may prefer the brilliance of larger standard-voltage lamps. As a general rule, plan for a light intensity in each outdoor area that makes it easy to do what you want to there without dazzling, producing glare, or bothering the neighbors. See also the notes on levels of brightness in Chapter One (page 8), and the specifications given in Chapter Three for the lighting of different garden areas.

Color rendering is the effect of a lamp or light source on the colors of objects. Look also at the section on the use of color in outdoor lighting on page 43.

The requirement for auxiliary parts may make the fixtures more expensive and larger. Some lamps will fit almost any fixture, while others may require specific fixtures or an attached ballast.

Lamp durability determines the frequency of some maintenance procedures. How much you worry about the rated life of lamps depends on how easy it will be to change them (is the fixture easy to get to, or at the top of a tall tree?), and how much they will be used. For lighting systems that will be on hundreds of hours a year, or for lamps in difficult spots, it may be best to use longer-lived bulbs. Numbers are manufacturer's rated lives.

Incandescent: Mini-lights have very long life. Other bulbs are generally not long-lived, lasting from 1000 to 3000 hours. Krypton PAR lamps are rated for 4000 hours.

Lamp costs range from the smallest incandescent lamps, which are least expensive, to the high-intensity discharge lamps, which cost considerably more. The cost of the lamps may be most important if your system is a large one. Bear in mind, too, that the more expensive high-energy lamps may last much longer than less-expensive incandescents.

You can see that the best lamps for most home outdoor lighting are standard- and low-voltage incandescent lamps and mercury vapor lamps. Incandescent lamps come in sizes and shapes to fit most needs, and have a pleasant, warm-white color rendering. Mercury vapor lamps are very energy-efficient. They have an unusual color rendering, but many people find this attractive. Fluorescent, quartz incandescent, high-pressure sodium, and metal halide lamps are available in fewer sizes, and the sodium and metal halide lamps have color rendering that many people find unattractive, but they can be useful for some specific lighting jobs.

If the main consideration is efficiency, mercury vapor or high-pressure sodium lamps will be your choice, provided you are comfortable with the color rendering these give. If color rendering similar to daylight is important, incandescent lamps will be most acceptable. Quartz incandescent lamps can provide brilliant illumination while giving light in the incandescent color spectrum, and they can be installed on circuits with dimmer switches. If you have to light a large area brightly, high-pressure sodium or metal halide lamps can do this more cheaply than quartz incandescent lamps, but they are too bright for many situations and cannot be inexpensively dimmed.

When choosing your lamps, consider the characteristics discussed in the table (page 29) and become familiar with the brightness and color rendering of each type by visiting lighted gardens, parks, and public places in your area. You can learn much about the characteristics of mercury vapor light, for instance, by studying the visual effects of streetlights, many of which are of this type (look for those giving off a blue-white light). Visit friends who have outdoor lighting systems using standard- or low-voltage incandescent lamps. Metal halide and fluorescent lamps are less frequently used outdoors, but your lighting distributor may be able to tell you of local installations you can visit. Notice also your physical and emotional response to each type of light: would you like to sit quietly on your patio under this or that type? Does it make people look odd? Does it make you feel relaxed or energized? All of these questions are important in making your outdoor lighting useful and beautiful to you.

Lighting Fixtures

Outdoor lighting fixtures are available to do any job from uplighting a small tree to floodlighting very large areas. Most are designed for use with incandescent bulbs. Many are also available for use with tubular quartz or fluorescent bulbs, mercury vapor, high-pressure sodium, or metal halide lamps, incorporating the special fittings and built-in ballasts required.

The best way to choose your lighting fixtures is to first study the design principles and process in Chapter Three, taking a look at what you need to light and why. Next, decide on the specific lighting techniques you will use to do this, from the first part of this chapter. Finally, refer to the survey of lighting fixtures given below to make informed choices from the many products available on the market. See page 39 for a list of manufacturers if you can't find what you need from your local lighting supplier.

Above: This bullet fixture is hidden among the leaves during the day.
Right: Not all fixtures need to be hidden. Some, such as this "wrenhouse" fixture, are meant to be seen in the daytime.

Lampholders and Bullet Lights

A lampholder is a simple fixture, often just an electrical junction box attached to one or more sockets, with no shield or lens, and designed to hold floodlights and spotlights. Though they are inexpensive and frequently used, lampholders often do not provide attractive outdoor lighting, because they do nothing to shield the lamp or reduce glare.

Bullet lights are similar to lampholders and almost as common, but they improve on them by enclosing the lamp in a cone-shaped metal or plastic shield. This not only cuts glare, it also protects the lamp and socket from debris and moisture, and so avoids short circuits. Additional protection is provided by the plastic or glass lens sold with many bullet lights; this is essential if you are using R or ER type incandescent lamps outdoors, because moisture or sudden temperature changes can cause these indoor lamps to explode.

Lampholders and bullet lights are used for uplighting, downlighting, moonlighting, spotlighting, grazing light, and safety lighting. Silhouetting is often done with bullet lights trained upward on a wall behind a plant, and shadowing with bullet lights trained on a plant in front of a wall.

Bottom left: This well light will be installed in the ground, with the top flush with the soil surface. Bottom right: Low-growing plants help to blend a fixture into the landscape.

Well Lights

To better conceal the light source, some uplighting fixtures can be installed underground, with the top of the fixture flush with the ground surface. These are known as well lights, and can be useful for uplighting trees, shrubs, walls, and signs. Well lights are available in both standard and low voltage, in plastic and various metals, and many can be ordered with a grill to better hide the lamp.

A grill hides the lamp and helps keep debris out of this well light.

Post Lights and Bollard Lights

Post lights are among the most common outdoor lighting fixtures. They are frequently installed at a front entry walk or driveway to provide area lighting, illuminate the path for safety, and welcome visitors. Post lights are available in various heights from 3 feet to more than 8 feet tall, and in plastic, enameled metal, and redwood. Usually the lamp is enclosed in a globe or chimney at the top of the post; this should be made of a light-diffusing material such as frosted glass or white plastic to avoid glare, especially if the globe is near eye level.

Bollard lights are very similar to post lights but are usually about 3 feet in height and enclose the lamp not in a globe but behind plastic or glass panels within the post itself. They are available in many attractive styles in both metal and wood.

Above: Post lights make strong architectural statements in the daytime as well as at night. When they are near eye level, frosted globes or shields reduce glare.
Right: Post lights should be compatible with other architectural elements in the garden.

Porch Lights
You needn't restrict porch lights to your front porch; they can also be useful for area lighting on the patio, or safety lighting when hung on walls near dark paths. Porch lights are available in countless styles, to hang both from walls and ceilings. You can choose different shapes, sizes, and designs to suit any porch or outdoor area that has walls or a roof for mounting. Choose a fixture with diffusing glass or plastic globe or panels if it must be at eye level, or locate the fixture overhead so that only its light is seen.

Low Area Lights
Low area lights (also called path lights) usually stand 4 feet or less in height; the most common low area fixtures are the standard "mushroom" and "pagoda" lamps. The light from these fixtures is usually diffused through plastic panels, or directed toward the ground by a shield or louvers to avoid glare.

Low area lights are most used for path and safety lighting. They are also useful for contour lighting and area lighting of flat surfaces or low plantings such as lawns and ground cover. Because they are one of the most common types of outdoor fixtures, low area lights are available in a great many styles and materials, with much variation in the amount of shielding of the lamp and the size of the area lighted by one fixture.

Top left: Entry lights and wall lights can be attractive and effective without being distractingly bright.
Top right: Low area lights focus attention on ground covers and other low plantings.
Bottom: Position stair lights to avoid shadows on the steps.

Left: A flower fixture is part of the view in the daytime, a source of light at night. Below: Flowers or other plantings can camouflage fixtures.

Above: Dark colored fixtures blend into the shadows of taller plantings.

Stair lights can be built in during construction or remodeling.

Fluorescent fixtures provide even wall illumination.

Recessed Stair Lights
These fixtures can be built into a masonry wall or steps to illuminate the stair treads and risers or the landscape floor. Although their purpose is mainly to provide safety lighting, they also add a background of fill lighting for nearby focal points.

Other Fixtures for Accent Lighting
Accent lighting adds interest to a garden in many ways, and may use small flower-shaped area lights in planter beds, punched copper fixtures hung on patio walls, or strings of minilights draped over trees or shrubs. For very precise spotlighting of garden focal points, a sophisticated fixture called a framing optical projector can match the shape of the light beam to the object being lit so that no light spills beyond it.

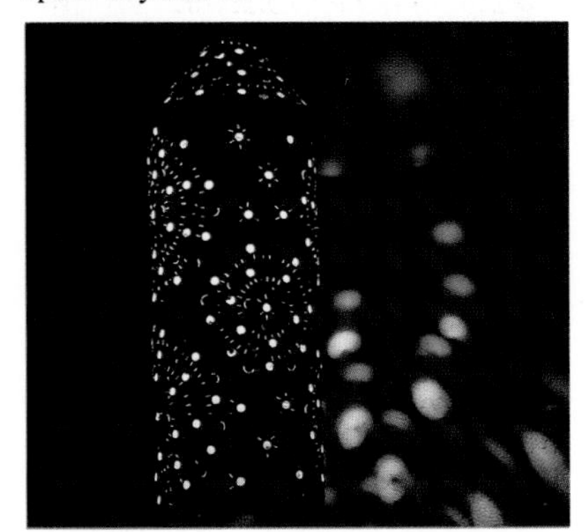

Light shining through the holes of this copper fixture creates an interesting pattern against a nearby wall.

Fixtures for Water Lighting
Swimming pool lights are of two types: wet-niche fixtures, which can be removed to change the lamp, and dry-niche fixtures, which require access to the back of the pool shell for re-lamping. Both use large lamps, usually of several hundred watts, and are built into the pool walls during its construction.

A number of fixtures are available for lighting small garden pools and other water elements in the garden. Nearly all can be wired to an underwater junction box during construction of the pool or fountain, or installed later, with the connection to power made above water level near the pool. Some of the fixtures are designed specifically as uplights, to be laid on the bottom of the pool, ideally at the base of a fountain, cascade, or other turbulent water to hide the light source. Others are simply sealed glass globes over incandescent bulbs that give a soft, all-around light. And some are mounted on stands to allow adjustment; these can direct light upward to light a fountain or be aligned in any other direction. Most of these fixtures for water lighting are available with lenses or globes in various colors, so you can try colored lighting here if you wish.

Fixtures for Fluorescent Lamps
Fluorescent light can give very even illumination of walls for silhouetting or grazing and can light a building facade or sign effectively. Fixtures designed for fluorescent tubes are usually long and narrow and can be positioned as uplights or downlights or to shine across a wall or ground surface.

Fixtures for High-Intensity Discharge Lamps
Several manufacturers make fixtures available in both standard models for incandescent lamps and with attached ballasts for high-intensity discharge lamps. When ordering, be sure to specify the fixture ballasted for the type of high-intensity lamp you'll use: each requires a different ballast. Some of the fixtures available for high-intensity lamps are bullet lights, post lights, and well lights. Most hardware stores carry high-intensity lights that use a high-pressure sodium or mercury vapor lamp.

Fixtures for Gas Lights
Gas lighting has been used for entry lighting and other outdoor fixtures in some parts of the country. The light it casts can be attractive and alive, but as the price of natural gas has risen, gas lights have become expensive to operate. Gas lighting is also an inefficient use of energy. Utilities can supply gas to existing outdoor gas lighting fixtures but are now prohibited by Federal regulations from making any new connections to gas lighting fixtures. For these reasons, only electric lighting systems are discussed in this book.

Sources of Lighting Materials

It's best to shop for fixtures and parts at your local lighting or hardware store, where you can inspect what you are buying. But outdoor lighting fixtures may be difficult to find in some areas. For this reason, we list several manufacturers of outdoor lighting fixtures and materials below. These manufacturers offer a wide range of products at various price levels. Some of them sell primarily on the wholesale level, while others will sell retail as well as wholesale. Several manufacture top-quality fixtures especially for the commercial market; while these may be appropriate in some residential situations, some homeowners may find them more expensive than they wish to pay. Others aim their products more to home lighting systems. These are often more economical, but there will be some variation in quality. Check for quality of design and materials, service available in case of problems, and guarantees.

Hydrel, Inc.
9415 Telfair Avenue
Sun Valley, CA 91352
Fixtures for downlighting, moonlighting, area lighting, uplighting, safety lighting, well lights for silhouetting, pool and fountain lighting. Fixtures for high-intensity discharge lamps.

Harvey Hubbell Inc.,
Lighting Division
Christiansburg, VA 24073
Many fixtures and accessories for downlighting, area lighting, moonlighting, uplighting, safety lighting, silhouetting, lighting of swimming and garden pools; also low-voltage fixtures and transformers, underground and underwater junction boxes, fixtures for high-intensity discharge sources.

Intermatic, Inc.
Intermatic Plaza
Spring Grove, IL 60081
Manufacturers of Intermatic controls and Malibu low-voltage lighting systems. Many lighting materials are available, including fixtures, wire, and transformers, and a wide variety of time clocks and controls.

Keene Corp.-Stonco
2345 Vauxhall Road
Union, NJ 07083
Many fixtures for area lighting, uplighting, and high-intensity discharge lamps. Direct burial boxes and other electrical materials are also available.

Kim Lighting
P.O. Box 1275
City of Industry, CA 91749
Fixtures for downlighting, area lighting, moonlighting, uplighting, safety lighting, silhouetting; large line of pool and fountain lighting fixtures. Many fixtures for high-intensity discharge lamps.

Leviton Manufacturing Company, Inc.
59-25 Little Neck Parkway
Little Neck, NY 11362
Manufactures switching systems useful for large lighting projects.

Lightolier, Inc.
346 Claremont Avenue
Jersey City, NJ 07305
Mainly a manufacturer of indoor lighting fixtures, also has many decorative porch lanterns, wall-mounted downlights.

Loran, Inc.
1705 East Colton
Redlands, CA 92373
Manufacturer of the "Nightscaping" line of low-voltage lighting fixtures and accessories. This includes many fixtures for downlighting, moonlighting, area lighting, uplighting, safety lighting, accent lighting, mini-lights, silhouetting and shadowing, and garden pool lighting. Also low-voltage transformers, time clocks, tree and deck mounts, lamps, wire, and plug-in switches.

Prescolite
1251 Doolittle Drive
San Leandro, CA 94577
Fixtures for area lighting, downlighting, uplighting, safety lighting, many porch lanterns, and wall-mounted fixtures for uplight, downlight, and grazing light. Fixtures for high-intensity lamps.

Sterner Lighting Systems, Inc.
Winsted, MN 55395
Commercial quality bollard fixtures for area lighting, for both incandescent and high-intensity lamps.

Sylvan Designs, Inc.
19767 Bahama Street
Northridge, CA 91324
Standard and low-voltage fixtures primarily in redwood, for area lighting, uplighting, safety lighting. Also micro-lights, transformers, low-voltage accessories.

Victor Manufacturing Company, Inc.
1020 Terminal Way
San Carlos, CA 94070
Redwood fixtures for area lighting, safety lighting.

Wellmade Metal Products, Inc.
860 81st Avenue
Oakland, CA 94621
Retrofit adapters for converting incandescent fixtures to fluorescent (see below).

Wendelighting
9068 Culver Blvd.
Culver City, CA 90230
Specializing in various models of framing optical projectors for the lighting of art works and fountains. Also some safety and area lighting fixtures, small accent lights.

Other Materials

Wire, connectors, switches, boxes, and low-voltage transformers link the lamps and lighting fixtures into a system. Conduit, fuses, circuit breakers, and the Ground Fault Interrupter (GFI) ensure electrical safety. Additional control and refinement are provided by dimmers, time clocks, and sensing devices. Retrofit adapters offer energy efficiency by converting very large systems from incandescent to fluorescent lamps.

Wire

"Wire" refers to a single strand of metal, and "cable" means two or more strands bound together. But in common usage, "wire" is also used as a generic term to mean either single or multiple strands.

Electrical wire is designated by type and size. The most useful type for outdoor wiring is Type UF (for Underground Feeder), usually used in a size number 12-2 (indicating a two-strand cable with each wire of size number 12). Some lighting designers prefer number 10 wire for its greater capacity (smaller numbers indicate thicker wire), especially for long runs. Most codes allow direct burial of UF cable without conduit; some may require conduit. Local codes also specify the required depth of burial. Some local codes require the use of wire rather than cable. The wire is usually type THHN or TW, and must usually be installed in conduit.

For low-voltage lighting systems, 12-2 UF cable is best. 14-2 cable is sold by some low-voltage manufacturers for use with their systems. You can use number 18-2 cable for low-voltage fixtures mounted in trees.

Wire Connectors

Good electrical connections are essential in outdoor wiring. Poorly made or improperly insulated connections may generate heat and cause a short circuit or a fire. In wiring your outdoor lighting, you will be making connections at each switch, receptacle, and fixture. The most common connectors are wire nuts, which come in several sizes, depending on the wire you are using. The wires are stripped of insulation at the end, twisted together, and threaded into the wire nut, which holds them tightly. For extra insurance, wrap a twist or two of plastic electrician's tape around the wires and the base of the wire nut. Always make the connection above ground, in a watertight box if a standard-voltage system, or in the way recommended by the manufacturer, if low voltage.

Some manufacturers of low-voltage lighting make special connecting hardware for use with their systems; this is often easy to use and relatively waterproof, and is worth ordering with the fixtures. Connections here should also be made above ground.

Conduit

Conduit is any pipe that contains and protects outdoor wiring, and is commonly required by the National Electrical Code on above-ground portions of 120-volt wiring. It may also be required below ground in your area, or its use may permit you to bury the UF cable more shallowly, if digging on your site is difficult. You may want to use conduit, in any case, to protect your wiring from damage by rodents or rocks in the soil.

The easiest conduit to use is plastic conduit made of polyvinyl chloride (PVC). PVC will not corrode, is easy to cut and fit, and is fairly inexpensive. Rigid conduit is steel pipe with threaded fittings. It is more expensive and more difficult to work with than PVC, but is extremely strong. Some local codes may permit shallower burial for rigid conduit than for PVC; if your site is hard to dig, you may wish to use rigid conduit.

Wire nuts come in various sizes. The color indicates the gauge of wire they fit.

Boxes, Switches, and Receptacles

Boxes contain the electrical connections made for switches, receptacles, or lighting fixtures. Depending on their purpose, they are available in several standard sizes and with various covers. Junction boxes, for containing electrical connections, usually have a solid cover. Other boxes are covered by switches or receptacles. Boxes for outdoor use have special gaskets and covers to make them weatherproof.

Many refinements have been developed in electrical switching devices. Dimmers (see below) can precisely control the level of illumination on each circuit. Control switches are available that can be plugged into any receptacle in the house to turn on specially wired outdoor circuits. Wireless sensing systems (see below) turn on any number of pre-wired circuits at the approach of a car or other object, and are equipped with an emergency switch that can manually activate all lighting immediately.

Outdoor receptacles can be especially useful for temporary lighting fixtures in game and entertainment areas, and can be used in other parts of the garden to make the lighting more flexible.

Ground Fault Interrupter

In a ground fault, a bare wire touches any grounded object, tool, or other conducting material. This redirects the current from its normal path; it may flow along the grounded object or through your body if you are standing on wet ground or in water. A Ground Fault Interrupter (GFI) protects you from this and is required by electrical codes on many outdoor wiring circuits (check your local code for requirements). The GFI monitors the flow of current to ground in the circuit into which it is wired, and under conditions of a ground fault will cut off power within about $\frac{1}{40}$ of a second, preventing serious harm to you.

There are three types of GFI. The first is built into a receptacle. This is easy to install, and will protect any receptacle or fixture further along the same circuit. For this reason, if this type of GFI is used outdoors, it is best wired in at the beginning of the circuit. The second type combines a circuit breaker with the GFI and is installed in the breaker or service panel. GFI's of this type are available with circuit breakers of 15, 20, 30, and 40-amp rating, for 120-volt 3-wire circuits. Plan to use this type if you will be running new circuits from the service panel. The third type of GFI is plugged into any receptacle. This will provide ground fault protection of tools or fixtures plugged into it.

Fuses and Circuit Breakers

It is essential that all outdoor lighting circuits be protected not only by a GFI, but also by a fuse or circuit breaker. The fuse will melt, or the circuit breaker will trip, if more electricity is drawn by the circuit than it is designed to handle. This protects you from the buildup of heat within the wires and the consequent melting of the insulation and potential for fire.

The number of lighting fixtures you plan to install determines whether you will be able to add on to an existing circuit or must add new circuits and perhaps even a second service panel. Check the fuse or circuit breaker rating on any existing circuits you plan to use, and calculate the additional watts of power you can safely draw from it. If you are adding circuits, design them to safely carry the amount of current needed. See page 70 for help with this calculation.

Plastic conduit (PVC) is glued with smooth (slip) joints, or threaded to attach to metal parts.

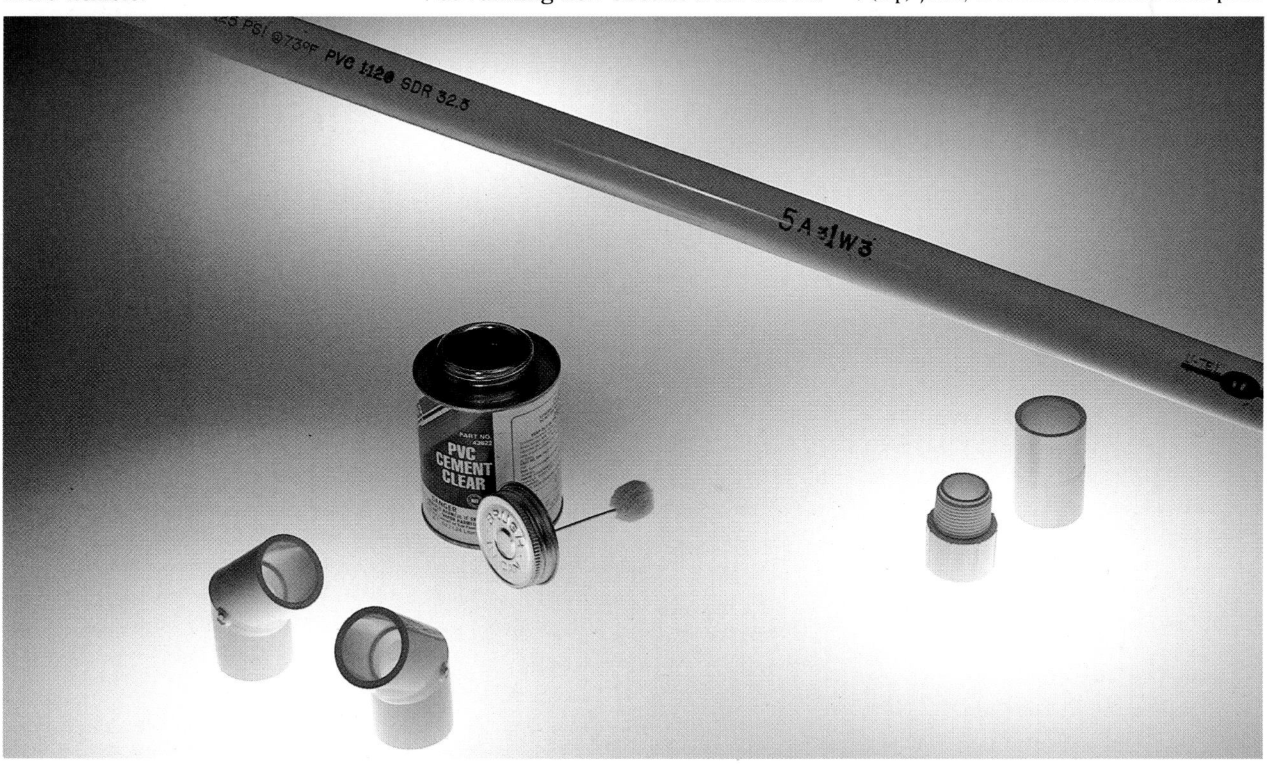

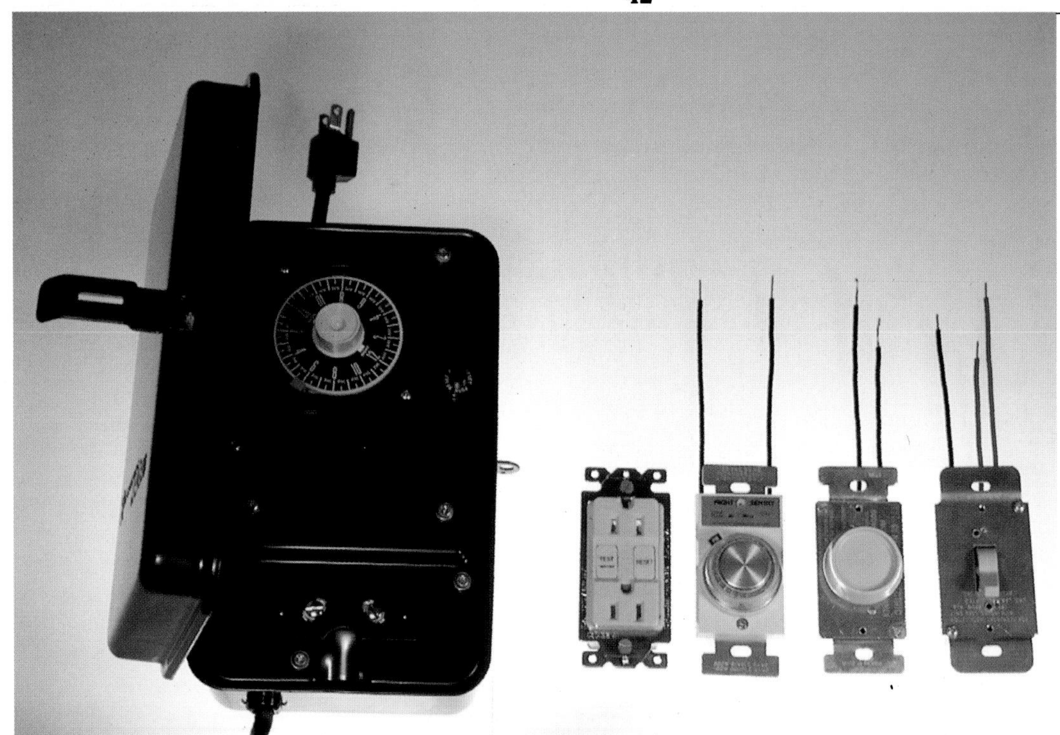

Power to lighting systems can be controlled by various switches. From left: A time clock, GFI, photocell switch, dimmer, and standard switch.

Low-Voltage Transformers

Low-voltage lighting systems require a step-down transformer to convert the incoming 120-volt power to 12 volts. These transformers are sold by most low-voltage manufacturers. The best are "raintight," for outdoor installation, have a built-in grounded barrier between the primary and secondary power leads, and can be plugged into an outdoor receptacle. If the receptacle does not have a switch operating it from indoors, plan to install one. Order the transformer sized for your lighting needs (see page 70 for information on calculating the wattage of the system). Some transformers are available with a photocell to turn your outdoor lighting system on at dark and off at dawn, or with a built-in time clock.

Dimmers

Dimmer switches are not required equipment on outdoor lighting systems, but they can be extremely useful. Dimmers are usually used in place of standard switches, allowing precise adjustment of the light output from incandescent lamps. They are also available for fluorescent lamps, but those for mercury vapor and other high-intensity discharge lamps are too expensive for most home uses. As we'll discuss later, the use of multiple switches and dimmers on each of your outdoor lighting circuits greatly increases the versatility of your lighting, making many different combinations and effects possible. Dimmers can also increase lamp life by limiting the wear on lamp filaments.

When buying dimmers, choose one with at least 25 per cent more capacity (in watts) than the electrical load on the circuit, and get one of good quality, which will last much longer.

Time Clocks

A time clock allows you to program when and for how long your lighting system is turned on. These clocks are not as flexible as those available for landscape sprinkler systems, and only allow you to set one "on" and one "off" time; the clock then turns the system on and off at these times every day. Some models can be set to slightly vary your time settings from day to day. A time clock can be an important part of a security lighting system, since it can give the impression that someone is home to turn on the lights. Some manufacturers combine the time clock and transformer in a single weathertight case.

Sensing Devices

A photocell is a light-activated sensor that can operate an electrical circuit automatically to turn on or off lights, fountains, or other electrical equipment. Photocells are widely used on city streetlights, and can be useful in the home garden when wired to a low-voltage transformer, serving as a sort of time clock.

Another type of sensing device is the infrared sensor used in sophisticated lighting systems. When a person crosses the infrared beam, the sensor turns on a set number of lighting fixtures. This system can be wired to turn the lights off again after a certain time has elapsed. Such sensors can be useful in security systems and even in path lighting, continually activating the lighting ahead of a person.

A recent innovation is the wireless sensing device used in some security systems. One version of this is a metal detector that is activated by a car or other object entering a property, and by means of a radio signal turns on any lighting circuits wired to receive it. For a sophisticated security lighting system, this can be combined with a burglar alarm of some type, and with an emergency "panic" switch that turns on all lights.

Retrofit Adapters for Energy Saving

One additional item that deserves mentioning is the retrofit adapter for converting lighting systems from incandescent light to the more energy-efficient fluorescent light. Available from several sources, they may make a important economic difference only in large lighting systems, as in multi-unit apartments.

THE USE OF COLOR IN OUTDOOR LIGHTING

Color is difficult to use well in outdoor lighting. It is a powerful tool and can be overpowering in its effect on the appearance and feeling of the nighttime garden. Here are some guidelines for its proper use.

One reason that colored garden lighting is so powerful is that colors have definite visual and psychological effects. The garden looks different, and people feel different, in environments tinted by various colors of light. "Warm" light sources (those with high levels of red, orange, and yellow light) may make people feel more comfortable in the garden. "Cool" sources (high in blue and green light) can feel slightly eerie, although they also give the garden an exciting air of mystery and spaciousness. Each color of light also has particular color rendering characteristics (see page 31), making some plants and objects look better, and others worse, in the garden. Warm white light, such as that from most incandescent lamps and warm white or daylight-type fluorescent tubes, renders plants, people, food, and most garden surfaces much as the sun does in daytime, so may look most "natural" to people. Blue-white and blue light bring out the green of foliage but make the human complexion and food served outdoors look unnatural. Yellow and amber light make green plants look bleached or sick. Red looks unnatural and hot. Pale pink light can make people look healthy and vibrant and can give a look of life to white marble statuary, but is not attractive on plant foliage.

There are several ways to add color to your lighting. The simplest way doesn't involve any special lenses or filters, but is done by mixing different lamp types and intensities. When using incandescent light, lower wattage lamps, especially those 40 watts and less in size, are warmer (having more yellow and orange) in color, while high-wattage and quartz incandescent lamps give a whiter light. Low-voltage incandescent lamps, because they have a different type of filament, are also more white in color. The blue-white of fluorescent lamps, the strong blue of mercury vapor lamps, and the slight greenish tinge of metal halide lamps, can all be distinctive when used to mark focal points in the landscape. Mercury vapor lamps, in fact, are used by some designers to "moonlight" entire gardens, creating—according to your point of view—an eerie or a magical quality to the landscape. If you do choose to mix lamp types in your lighting for effect, do it carefully. Use one type of lamp as the primary light source, adding others only for a specific reason, as to mark an important focal point, a special plant, or the house entry. If your garden is small, combining several lamp types can be confusing, and may be best avoided.

Another way to get color in your night garden is to install special tinted floodlights or lenses in the lighting fixtures. This is the most commonly used technique in colored lighting, but it should be used carefully if the effects are not to be too strong. Don't simply arrange multicolored lights alternately in a row, or place a brilliantly colored lamp beneath a nondescript shrub. Use color instead to bring out what's important in the garden, in a carefully planned way. Use mainly white light, saving the more intense colors for occasional contrast or accent. This will give clear information about what is important to look at, and avoid giving confusing messages. Because colored light is a powerful tool, use it carefully.

The liberal use of colored lights gives Butchart Gardens a magical, storybook appearance at night.

Planning Your Lighting

Design a lighting system to meet your family's needs.

A landscape designer works with simple design concepts and the raw materials of trees, shrubs, flowers, decks, walkways, and existing buildings to create a beautiful garden. In much the same way, you can combine the design concepts of outdoor lighting with a few fixtures and electrical parts to create lighting magic in your garden.

In the first two chapters you learned the basics of outdoor lighting. In this chapter you'll use these guidelines to design a system that meets your lighting needs and beautifies your garden. By designing your own lighting system, you can make your garden exactly serve your family's needs. The ideas and examples in this chapter will help you formulate your own ideas and techniques for bringing out the beauty of your garden at night.

Even if you plan to have a designer or contractor do your lighting, the design principles and checklists in this chapter will help you to understand and supervise the work. If your budget is tight, you can develop your lighting plan now with the information provided here, and install your system later as finances permit. You can also design your system so that installation can be completed in sections rather than all at once. Fixtures can be installed first in high-use areas, such as entryways or patios, and later in remaining areas.

Whatever your situation, the design principles and ideas in this chapter will help you to bring your lighting dreams and plans into existence.

Opposite: A walk in this garden is like a walk in the woods on a moonlit night.
Right: Lighting emphasizes the architectural features of this entryway.

DESIGN PRINCIPLES

Certain principles can help an outdoor lighting system to be both useful and beautiful, while avoiding troubles and annoyances. These are simple points to keep in mind while you're designing a system. The same points can serve as a checklist when reviewing somebody else's design to see if it will serve your needs.

Study Natural Light

See how the sun highlights plants and landscapes, the pattern of shadows on a wooded hill, silhouettes of trees against the sunset sky, the sudden brilliant beam of sunlight escaping from behind a cloud. The sun spotlights, accents, downlights, reflects from water, and grazes a wall or fence—our interior and garden lighting systems are only imitations of what the sun does so beautifully. And notice the dramatic difference in feeling and mood between daytime and moonlight. Especially notice the lighting effects provided by the full moon, the gentle glow and the soft shadows as the light filters through a tree or illuminates a garden. In garden lighting as in any kind of design, the effect will seem most "natural" when it imitates nature.

Don't Overlight

As lighting designer William Lam suggests, a small quantity of the right kind of light can produce much better visibility than large quantities of an inappropriate kind. Less light can also create more subtly beautiful effects. This means carefully deciding where light is needed, using the appropriate lamps and fixtures in each area, and avoiding glare. Use lighting techniques, fixtures, and lamps that are in scale with your garden. Choose low-voltage or small standard-voltage fixtures for small gardens, larger lamps and fixtures in large landscapes.

Consider the Reflectance of the Plants and Materials in the Garden

Every patio, fence, or plant has its own level of reflectance, depending on its color and texture. The light reflectance of a plant is also affected by the openness or tightness of its form. Plants and materials with high reflectance are brilliant when lit. Plants and materials that absorb more light require brighter lighting to stand out.

Below: Maple leaves downlit by the sun make a bright tracery against a dark background.

Rely Primarily On One Type of Lamp

Chapter Two compared many different types of lamps. Each has a particular color rendering and range of intensity; decide which is the most appropriate type of lamp for your system, and use it as your mainstay. In most cases this will be standard or low-voltage incandescent lamps, or mercury vapor for certain types of "moonlighting" effects. Occasionally it may be appropriate to use a few lamps of a contrasting type to emphasize a special feature or focal point in the garden, illuminate the main house entry, or give other important visual information. Fluorescent, high-pressure sodium, and metal halide lamps are usually best used for specific purposes in the home garden.

Use Light to Show What's Important in the Garden

Illuminate focal points and safety hazards more brightly than less important areas. Arrange the lighting to show visitors where to enter the driveway, in which direction a path leads, or where to congregate on the patio. Use contrast between brighter and dimmer areas to give these clues, rather than brilliant spotlights.

Left: Natural backlighting makes leaves translucent and creates strong shadows.
Above right: The sunset silhouettes a giant saguaro.

Moonlighting and accent lighting combine to bring out the charm of this southern garden.

Hide the Source of Light

Light looks most natural when its source is hidden. Choose locations that hide or shield lighting fixtures so the light is seen and not the lamp. The main reason for hiding the light source is to avoid glare. Glare is brilliant light that distracts us from seeing what we need and want to see in the environment. Glare is what you experience when a bright light shines in your eyes. In outdoor lighting, it is usually caused by visible, bright lamps or lighting fixtures. There are several ways to conceal lighting fixtures and avoid glare. You can position the light fixture and lamp high above the ground, you can use several smaller light sources instead of one large one, or you can attach a diffusing lens or grill to the fixture. You can also position the fixture to diffuse its light through the leaves of a tree as is frequently done in moonlighting, or aim the light toward a light-colored surface such as a painted building wall, so that it illuminates by reflection.

Use Several Different Lighting Techniques

Your lighting will be more attractive if you combine several techniques in the garden. You might, for instance, use moonlighting from tall trees as the main light source, then add occasional accent lighting from low fixtures and decorative area lighting to mark the entry. At the same time, avoid too much variety: use primarily one type of lamp, match the style of fixtures where possible, and combine the lighting effects harmoniously.

Keep Fixtures Out of the Way

Choose all fixture locations carefully, so they won't interfere with mowing of lawns, walking, or plant growth. This caution is especially true with 120-volt systems; the fixtures may be very difficult to move later. If you expect to relocate some 120-volt fixtures as the plants grow or your garden lighting needs change, order them with a portable stake mount.

Plan Your New Garden for Lighting

New gardens of young plants are hard to light. Small container shrubs and trees provide no architectural grandeur for uplighting or moonlighting. Sparse perennials in a new flower bed are hard to light as a garden accent. If your garden is young or even new, place a few larger specimen plants in the scheme to have some subjects for lighting. Use smaller lamps for younger plants, changing them to larger lamps and relocating the fixtures as the plants grow.

Make Your Lighting System Adjustable

Plan your system with several circuits, each with its own switch. Use dimmer switches whenever it's practical to do so. This will allow you to turn on only the outdoor lights that you need for any activity and to precisely control their levels. For example, you can turn on underwater lighting in a swimming pool fully for swimming, or dim it to provide area lighting nearby for dining or entertaining.

Put Lighting Controls Indoors at a Central Location

One set of switches in a convenient, accessible place is all that's needed for most systems; this way you can orchestrate all your garden lighting effects easily from one location. For larger systems, you may want to install additional banks of switches in important use areas to control the nearby parts of the system. For instance, you can place switches near the swimming pool to operate the pool lighting and adjacent tree uplights, and other switches inside the garage to operate the driveway lights. You should be able to control each of these areas from the main switches indoors as well.

Work With Electrical Safety in Mind

Learn about the local electrical code from your contractor or building inspector, and then design your system to comply with the code's requirements. Have your lighting plan checked by a licensed electrician or designer, especially where fixtures or wiring will be near water. Hire a contractor to do any work you're not comfortable doing. Get all required permits for the system installation.

IDEAS FOR LIGHTING

Here are specific suggestions that will make your garden lighting useful and attractive. Each area and element in your garden is individual; certain techniques, fixtures, and lamps will work best in some situations, yet won't work in others. The level of illumination needed in each area also differs. If you've wondered how best to light a brick patio, a wooded driveway, or a specimen oak, the answers are here, based on the recommendations of many lighting designers. These suggestions can inspire your imagination as you work through the design process (beginning on page 65), and provide answers to your lighting problems.

When the recommendations refer to "bright" light, remember that this is defined (in Chapter One) as light of 8 footcandles or more; "medium-bright" light is light of .5 to 8 footcandles, and "dim" light is light of .5 footcandle or less. What is most important in any lighting scheme is the relative brightness of different areas, not the absolute brightness of each area. Use fill lighting as a background and bright, focal light for contrast.

Lighting the Driveway and Garage

The lighting at the entrance to your property should convey a welcome to visitors and show them how to enter your property. If your street is lit by streetlights, you may not need much supplemental lighting. If there are no streetlights, provide some soft light for the driveway. Use different intensities of light to direct traffic. Point out the entrances to driveways and walks with brighter lights, and use softer lighting along the rest of the driveway or walk. Avoid bare lamps at eye level and other sources of glare. Moonlighting is often used here to shed soft, diffused light from trees or structures high above (see page 19). If you can't light from high overhead, use either low area lights less than three feet high or overhead fixtures taller than 8 feet. Choose a fixture that shields the lamp from view or encloses it in a diffusing globe to prevent glare. Moonlighting can make a long, wooded driveway spectacular, especially if low path lights illuminate attractive accent plantings or critical corners along the way.

Lighting at the garage and outdoor parking areas is critical to the safety and security of many homes. Uniform general lighting of this area can be done with 40, 75, or 100-watt incandescent floodlights, or if the area is large, with mercury vapor lamps of 75 watts or less. Use fixtures with good shielding of the lamp, and hide them by mounting them in bordering trees or high on surrounding buildings. To light the front of your garage, you can use 75 or 100-watt Type A incandescent lamps (or the energy-saving 70 and 95-watt) enclosed in diffusing fixtures and mounted just above eye level on either side, so that they illuminate the inside of the car.

This garage light provides uniform general lighting as well as safety and security lighting.

1701

The House Entry

Your lighting should clearly mark the entry to your house. If you use mercury vapor moonlighting in the rest of the front yard, you can light the front porch with the more yellow light of incandescent lamps or simply provide slightly brighter light at the porch. To avoid glare, use reflecting or diffusing fixtures near the front door. On the porch you can combine low-wattage incandescent lamps in wall-mounted porch lights near the door with brighter, recessed lighting overhead. This will illuminate the faces of night visitors but avoid blinding them.

If you have a two-story house with an attractive facade, you can create a dramatic effect by uplighting it. Use fixtures several feet away that graze the facade with light. Use ER or PAR 75-watt incandescent floodlights; their light will illuminate the house front and eaves, the porch, and nearby tall trees. Fixtures for ER or R lamps should protect the lamp from moisture with a lens.

You may find it hard to create attractive front yard lighting because of the glare of a nearby streetlight. In some cases local power companies will install a shield in the streetlight to protect you from this.

Ideas for Security Lighting

It isn't necessary for security lighting to be harsh or glaring. It doesn't even need to be extremely bright. It is possible to design attractive garden illumination that puts light for safety on main paths, congregating areas, and hazardous areas, and lights for security on the porch area, house corners, low windows, side and back doors, and any spots where intruders might enter the property. This kind of lighting provides bright light where it's needed, accents garden focal points, and leaves some areas lighted less brightly for contrast. You can do it with moonlighting fixtures placed high in trees and aiming back toward the house (while avoiding shining it into windows). See the notes on the lighting of trees, page 56, for lamp recommendations. Or it can be done with low area fixtures using small incandescent lamps positioned to light the landscape floor clearly. You can also install a bright 250-watt quartz floodlight as a backup for the other lighting, to be turned on in case of emergency. If you wish to use a high-intensity yard light, try to position it to avoid glare and turn it off when it's not needed.

Your security lighting may be a part of larger systems that light entire areas of your yard, or it may be a separate system, especially if your property is large. In either case, the switch should be indoors for your own protection. If it is a separate system and you expect to operate it for many hours each night, you may wish to use mercury vapor, metal halide, or high-pressure sodium lamps for their energy efficiency. If you do, use them sparingly so that the bright light and distinct color rendering don't ruin the effect of your other lighting, and use the lowest wattage lamps that allow you to feel secure. For example, 30-watt high-pressure sodium lamps are now available, and give ample light for most home security systems.

If you are using any of these high-intensity discharge lamps for your security lighting, you should also include some quartz incandescent lamps in your outdoor lighting system, because the high-intensity lamps take a few minutes to warm up, so they don't provide immediate light when they are turned on.

Many types of switches, sensors, and alarms are available to add to a security lighting system, and these are what make it truly effective. Plug-in switches can be useful indoors, especially in the kitchen, front hall, or in whatever room you spend time in the evening. With these you can switch on any circuit that has been wired to receive their signal, from anywhere they're plugged in. Time clocks, especially those with a variable program, make your house appear occupied even when you're out. And several types of sensors, including wireless sensors that activate your lights whenever someone enters the property, can be helpful. Tailor the switching of your security lighting to your own needs, and install what makes you feel secure in your home.

Opposite: This entry lighting leads guests to the door.
Right: A recessed lamp in this overhead fixture avoids glare in the eyes of arriving guests.

Safety Lighting

The goal of safety lighting is to make you and your guests feel comfortable and move safely in the garden. To do this, you need to illuminate outdoor stairs and steps, uneven ground surfaces, water edges, and any other hazards. Safety lighting should also help direct traffic in the garden by lighting paths and leading people from one point to another. Light heavily used paths and those with an uneven surface most brightly. Position fixtures to light both the treads and risers of stairs, and where they won't be blocked by a person's shadow.

If you have good general lighting in your garden, you may not need to add special fixtures for safety lighting. Moonlighting, for example, can provide all the safety lighting you need. If you don't have the large trees or tall buildings necessary to do moonlighting, provide safety lighting with path lights and other low area lighting fixtures. "Mushroom" and "pagoda" fixtures are commonly used, often with 30 or 40-watt standard-voltage incandescent lamps or smaller low-voltage lamps. You can also use small aluminum or bronze flower-form fixtures, bollard lights, or area lights on posts. For paths and steps bordered by walls, build in recessed stair lights. Safety lighting should illuminate an area clearly, but avoid glare. Check stair fixtures from the bottom of the stairs to be sure they're well shielded, and order post and bollard lights with a diffusing globe or lens.

Check the manufacturer's specifications for the recommended spacing of each fixture, but don't follow this slavishly. Place fixtures first where you know you want light: at stairs, path corners and intersections, and any spots you want to accent. Place other fixtures to dimly light the entire path. Some irregularity in the spacing can be pleasing.

The bright lighting on this patio invites active use.

Lighting Patios, Decks, and Terraces

The lighting of your patio, deck, or terrace should be similar to that of your home's interior so that the interior is visually joined with the outdoors. It should also serve whatever activities you enjoy here: outdoor dining, games, conversation, or parties. Because you will often view your garden from the deck or patio, this should be the center of your lighting scheme, with attractive views of the night garden seen in every direction.

Avoid area lighting that gives everything a flat, dull look. Combine floodlights or diffused area fixtures with accent lighting of nearby garden features, reflected lighting from a painted house wall, or background lighting of tall shrubs.

If you are lighting your garden primarily with mercury vapor "moonlighting," you should still consider using incandescent lamps on the patio, because they make faces and food look warmer and more natural.

Flexibility is important in lighting a deck or patio. Provide separate switches for large area lights and small focal or accent lights. If the switch is a dimmer, you can make the lighting brighter or softer according to your mood and the occasion.

One designer suggests that you not light areas for outdoor tables too brightly, leaving room for the special accent of kerosene lamps, candles, or other non-electric party lighting there. Also install plenty of convenience outlets for temporary electric lighting such as low-wattage lamps in Japanese lanterns (powered by an extension cord). The outlets can also serve small radios or other electric appliances. See page 60 for more ideas for special-occasion lighting.

Lighting of wood decks is similar to that of any patio. Special fixtures are available from some manufacturers for installation under deck benches, or to allow you to attach small fixtures directly to the wood surface.

Although the switches for your garden lighting should be indoors, you may wish to have a master control panel on the patio too, to orchestrate the lighting effects during outdoor parties or entertainments.

Work, Utility, and Storage Areas

These active outdoor areas need bright illumination, whether you are lighting your potting shed, tool storage area, or garbage cans. Think about what you do in the work area, and whether the focus of the light should be on the ground level or on a raised bench or table. For storage areas, decide whether you need light inside as well as above the door, and position the fixtures to illuminate all the corners. Fixtures and lamps to use for outdoor work areas include PAR incandescent floodlights of 75 and 150-watt sizes (or the energy-saving equivalents of 65 and 120 watts), hung overhead in shielded fixtures, and some additional small spotlights (including low-voltage fixtures) that put light on the specific work surfaces. Don't forget to light the path from the house to the work area. You can use conventional indoor bulbs and fixtures for any enclosed or covered storage areas.

In a small yard, work area lights may also serve as lighting for the patio or other areas. If so, fit the system with a dimmer, so that the lighting intensity can be varied.

Lighting for Recreation

Badminton, volleyball, croquet, and other games can be played in the garden at night if lighting is adequate. These games require bright, uniform lighting without glare. Install recreational area lighting on a separate circuit because of its high power demands and intermittent use.

The most useful lamps for game lighting are PAR incandescent floodlights of 75 or 150 watts, or quartz incandescent lamps in a 90-watt PAR shape. For extremely bright light, use a 250 or 500-watt tubular quartz lamp in a reflecting fixture. If the lamps will be on for more than 1000 hours a year (several hours a night), consider also the more efficient and long-lived mercury vapor lamps. Install all lamps in well-shielded fixtures, hung high overhead to keep light out of players' eyes. The least expensive mounting method is to hang them on a nearby building, but if this is not possible you can set tall wood or metal posts in the ground where they're needed.

If you're providing light for a lawn game such as croquet or a putting green, you need bright lighting of the entire game area from the center sidelines. For net games such as badminton or volleyball, also position the floodlights at the center sidelines, but aim one at the net and another two at the rest of the court. An archery or darts target should be lit from behind the player, and some light also should be aimed beyond the target to help you find lost darts or arrows.

Because game lighting is usually very bright, be especially careful not to aim it into neighbors' yards, and turn it off when you don't need it.

Lighting extends the time a gardener can pot or propagate plants in this lath house. Placing the floodlight well above the work area provides general illumination without creating glare.

Lighting Water in the Garden

The most important consideration in lighting garden pools, streams, and fountains is safety. The presence of electricity near water offers great potential danger, and you should not do your own wiring or lighting here unless you are quite sure you know your local code's requirements and can do it safely. The code will probably dictate how close to water you can place any wiring, and it may require a GFI on any receptacle or fixture within 10 feet of the water's edge. Get all required plan approvals and inspections, and have your final installation checked by an electrician before the power is connected.

Direct downlighting of water is not usually attractive. It makes the water surface opaque and ruins any reflection of nearby landscaping. You can use the water as a reflecting pool simply by leaving the surface dark and lighting surrounding landscaping, and you can add supplemental underwater lighting that can be turned on separately. Underwater lighting won't be attractive, however, if the pool is cloudy with sediment or algae, so should only be installed if you can keep the water clear.

When lighting a garden pool, hide the light source. The best way to do this is to build in a small cave or grotto at one end when making the pool. You can then put the fixture inside this grotto so that only reflected light is seen. You can also build in a recessed "wet-niche" swimming pool light (see page 38) to the wall at one end of the pool; this should be equipped with a low-wattage lamp (under 40 watts of incandescent light for most garden pools), and positioned at the end from which the pool is viewed, to avoid glare. If you're lighting an existing pool, an underwater fixture can be set on the bottom amid large stones to hide it.

If your garden pool includes a fountain or cascade, place the fixture on the floor of the pool and aim it to uplight the moving water. Turbulent water reflects light well, sparkling as it moves. If the fountain or cascade is large, the turbulence may adequately hide a fixture placed directly beneath it. For a showy effect you can light a large fountain with a color wheel that projects a spotlight of ever-changing color upward from the fountain base. Some kits even combine these with music, but these effects will only work well in a very large, ornate garden; elsewhere they will dominate all other lighting effects and give the scene a circus atmosphere. See the notes on use of colored lighting on page 43.

Many fixtures are available for lighting garden pools and fountains. See page 38 for a discussion of them. They are of two major types: those that are installed at the time the pool is built, wired to an underwater junction box and recessed in the pool floor or wall; and those that can be added to existing pools and plugged into receptacles near the edge of the pool. When installing fixtures in an existing pool, hide the cord with trailing plants set along the water edge and with sand, rocks, or gravel where it runs across the pool floor. Check your local code to see how close to the pool's edge you

Top: The source of underwater light can be hidden by putting it on the side of the pool nearest the viewer.
Bottom: Water lilies are interesting when silhouetted with underwater lighting.

can place the receptacle. Both low-voltage and standard-voltage fixtures are available for garden pool lighting; the low-voltage fixtures are safest to install and use.

A stream is a special garden feature and can be lit from the side by small accent lighting fixtures. Hide low-voltage light fixtures in nearby grass, rushes, or wildflowers, and direct their light across the water surface to show the movement and sparkle.

Swimming pool lighting is usually installed as the pool is built, and is best left to the pool builder because of its complexity and potential danger. The usual fixtures are floodlights, often of several hundred watts, set into the ends or sides of the pool. "Wet-niche" fixtures can be removed from the pool for changing lamps; "dry-niche" fixtures are permanently installed and require access to the back of the pool wall to change lamps.

Check with the pool designer to be sure that the pool lights will be placed to avoid glare. As in garden pools, if only a single fixture is used it should be located on the end or side from which you usually view the pool, shining away from you. Be sure that the pool lights are on their own circuit, because the large lamps used have high power requirements. The pool circuit should be controlled by dimmer switches so that soft underwater light can give a pleasant background for patio entertaining and the lights can be brought up full for night swimming. If you use garden lighting nearby for mirror lighting or accent, be sure not to light the surface of the pool; this will destroy the reflection.

Spas and hot tubs in the garden should usually be lighted softly to illuminate the steps to the spa and the water edges for safety and to give the area a soft glow. Dim spill lighting or "moonlighting" from nearby garden areas is usually plenty of light for a spa or hot tub; you may also want to use low path or area lighting fixtures along the path from the house. Do not install any fixtures in hot tubs or spas except underwater fixtures approved for swimming pools.

Top: The sparkle of light from this waterfall is the "play of brilliants" that makes this garden come alive at night.
Bottom: Light from the pool casts an aquatic spell on this wall.

Illuminating Landscape Plants

The skillful lighting of specimen trees, groves, shrubs, ground covers, and flowers makes the garden exciting and magical at night. You can use many techniques to bring out the beauty of plants, including uplighting, moonlighting, spotlighting, accent lighting, grazing, silhouetting, and shadowing. In fact, if you design your lighting properly, the illumination you give the plants in your garden can also provide all the light you need for the front entry, the driveway, entertainment and patio areas, walks, and security, and do it beautifully. Lighting designers who do "moonlighting" of gardens have long followed this strategy, downlighting trees in such a way that beautiful patterns of light and shadow are cast across a patio, or a soft glow washes the entry. You can do this on a smaller scale too, spotlighting a flowerbed to spill over and light a path or silhouetting a plant in front of a wall to light the patio with reflected light.

Start with any large trees on your property, designing their lighting to emphasize texture, branching pattern, shape, color, and any seasonal changes they display. See Step 5 of the design process (page 68), and page 91 for suggestions. Once you've decided how to light the trees and how far the light will spill into other areas, plan lighting for the shrubs, lawn areas, ground covers, low accent plants, and flowers. Consider the same characteristics you did for trees. Does the plant have striking flowers? What about berries, or the pattern of bare winter branches in the snow? The lighting you plan should bring out the best each plant has to offer, while blending all garden plants and areas in the larger, unified lighting scheme.

Trees, especially tall trees, are perfect candidates for moonlighting. This striking technique requires downlighting from high in the branches. This will obviously require you to climb the trees both to experiment with temporary fixtures (see page 68) and to position the permanent fixtures. Learning to aim the downlights to diffuse the light through leaves, cast attractive shadows on the ground or buildings below, and illuminate those garden areas that need it, is an art that will require you to spend some time hanging fixtures, testing them at night, and repositioning them until you get it right. The guidelines given below should help you in this.

Moonlighting requires large, mature trees at least twenty feet high. Conifers are rarely used because of their great density, although some pines can become good subjects as they open up with age. Broad-leaved shade trees work best for this technique, especially those with an attractive branching structure and some bark texture, such as oak. If you do not have trees large enough for moonlighting but do have some patience, plant several now. See the list of suggested species on page 91, or choose any attractive shade tree that does well in your area and has a medium-to-fast growth rate. Plant the largest specimen size you can afford, and plan to begin lighting it in about ten years.

To simulate natural moonlight, hang the light source as high in the tree as possible. Light any tree with at least two lights, and up to eight or more for very large trees. Aim the fixtures so that the light beams cross, illuminating interesting branches and bark, and also the important areas below, such as the house entry, patio, walk, or garden steps. Cross lighting in this way gives the most attractive patterns of light and soft shadow below. Include some uplights too, positioned at least ten feet above the ground to avoid glare problems. These will illuminate the foliage above.

To moonlight a grove of trees, use the same techniques, but alternate the direction in which the light is thrown in alternate trees. This casts interesting shadows on the ground and highlights some areas. For an added sense of perspective, illuminate more brightly a "clearing" some distance into the grove.

Fixtures used for moonlighting are the common "bullet lights" for uplighting or downlighting, fitted with the longest shield available. For extra long shields, get your local sheet metal shop to make them for you. The longer the shield, the more narrow the beam of light, and the less glare. Incandescent lamps used in these fixtures are R, ER, or PAR lamps in 30-watt, 65 or 75-watt, and 120 or 150-watt sizes, respectively. Use incandescent spotlights or 90-watt quartz incandescent lamps in extremely tall trees.

New low-voltage lamps are also available for moonlighting and other tree lighting, including powerful quartz lamps and PAR lamps in a variety of beam widths from a very narrow spot to a very wide floodlight.

If you want to moonlight with mercury vapor light, order fixtures with a ballast for clear mercury lamps of 75 to 400 watts. For most gardens, the smallest sizes of mercury lamps give ample light. These lamps are relatively expensive, but have a much longer life than incandescent lamps.

Moonlighting is not the only way to light landscape trees. You can also uplight them from well lights set into the ground near their base. For narrow trees such as Italian cypress, position the uplights about 18″ from the trunk. Broad shade trees such as oak, elm, and weeping willow can be uplit from 3 feet or more away from the trunk. Flowering trees can be uplit from several feet away also, to highlight the small branches and blossoms. Combine some downlighting with this type of uplighting, and use it on a few trees that you want to feature prominently in the night garden. Use PAR, ER, or R floodlights and spotlights in the standard voltage wattages of 75 and 150 watts (or the energy-saving equivalents of 65 and 120 watts), or use low-voltage PAR lamps. Use the spotlights where the light must be thrown a long distance to the tree top, floodlights for shorter, broader illumination. Often you can combine the two types of lamps for the best effects. Try also the 90-watt quartz PAR lamp mentioned above.

Dense conifer trees can also be uplit, but the best way to do this is to lightly graze the branch tips with light. They are often attractive in silhouette because of their dark color and density.

Small trees with interesting bark such as strawberry tree or crape myrtle can be attractive focal points in the

The striking framework of this oak tree is illuminated by uplighting and reflected in the pool.

night garden. Try grazing soft light on the trunk rather than blasting them with floodlights. Small uplighting or downlighting fixtures with 30 or 40-watt standard-voltage R lamps, or low-voltage PAR medium-width floodlights work well for this. Use a fixture with a protective lens for the R lamps.

If you are lighting trees or tall shrubs in an atrium, you can light them from above with small incandescent floodlights (R, ER, or PAR) in sizes from 30 to 75 watts, or with recessed lamps built into soffits if they are under an overhang. Plant glossy-leaved plants like weeping fig (*Ficus benjamina*), and position the light sources as high above as possible for the most attractive light play.

Shrubs can be lit using the techniques of uplighting, silhouetting, shadowing, grazing, or downlighting from trees. You can graze tall, dense shrubs to form a background for perspective lighting or viewing from inside. Those with glossy leaves can be lit to reflect a soft glow onto a patio or deck. You can uplight shrubs having large, translucent leaves to make the leaves glow. Use small bullet lights and well lights for uplighting, or small bullet lights for downlighting, equipped with incandescent lamps of 30 or 40 watts. You can also use low-voltage uplighting fixtures with small spotlights, or small mercury vapor lamps.

Perennials and ground covers provide numerous possibilities for lighting. The best perennials and annual flowers for lighting are those with interesting or unusual foliage, such as plants with prominent leaves or flowering plants that hold their flowers well above their foliage.

For a dramatic effect, light sword-leaved perennials such as iris or fortnight lily by shadowing or silhouetting. If a flowerbed has a wall or fence behind it, you can silhouette all its flowers by washing the wall with light from fluorescent tubes or incandescent floodlights. You can also light flowerbeds with small, low-voltage border lights of about the same height as the flowers, or with slightly taller area lighting fixtures. Both of these use small A-type incandescent lamps.

These geraniums are a bright accent when lit from above.

Some ground covers, such as Japanese spurge, have an interesting surface texture. Well-shielded low path and area fixtures set near the planting or bullet lights grazing the plant surface with light can illuminate this. You can also use this type of fixture to light the edge of the ground cover where it meets a lawn or paving (see Contour Lighting, page 23).

For a listing of some common garden trees, shrubs, and perennials and some specific suggestions for their lighting, see the special section on page 91.

Other Garden Materials for Lighting

Lighting can bring out the attractive texture and color of many structural garden features at night. The outside house walls bordering the patio can be "washed" with light from recessed incandescent floodlights spaced evenly along the wall and positioned close to it. This is especially appropriate if you have used similar lighting indoors, because it unifies indoors and out. It is also an attractive, unobtrusive way to light a patio with reflected light, and can make it seem larger.

Outdoor areas roofed with wooden arbors can be downlit with fluorescent tubes concealed between double beams or joists. Fences, doors, and textured walls of wood or masonry are good spots to use grazing light. This brings out their attractive textures and by reflection can light nearby paths, steps or garden areas. You can even position grazing light fixtures at ground level to throw light across a textured garden floor such as an exposed aggregate concrete path or brick patio. For grazing light, use bullet lights designed for downlighting, fitted with incandescent floodlights or spotlights, according to the task. A narrow spotlight works best for tree trunks and other narrow objects, a medium floodlight for a door or low wall, a wide flood or several floodlights for the side of a building. Small R or ER lamps of 30 or 40 watts are usually adequate for grazing subjects.

A statue can be a central focal point for the night garden. Light it from above for the most natural shadows, and from several angles, avoiding washing it out with head-on floodlights. Use at least two fixtures and lamps, with a combination of floodlights and spotlights so light is not perfectly even; this will bring out its depth. Position the fixtures on the side from which the statue is usually viewed, and conceal them as well as possible to avoid glare. The best fixtures are bullet lights with deep shielding. Try standard voltage incandescent lamps of 30 or 40 watts (R or ER), or low-voltage quartz or PAR spotlights and floods. If you want to light a statue without throwing any light beyond it, investigate framing optical projectors. These are used to illuminate many fine art collections both indoors and out, and are described on page 38.

Above: Lighting contrasts the softness of this fern with the rugged texture of the redwood bark.
Right: A spotlit statue is a strong focal point in the night garden.

SPECIAL-OCCASION LIGHTING

A well designed outdoor lighting system can make your garden more beautiful and useful for everyday use. It can also make it a special, magical place for entertainments and holidays. Here are some suggestions for lighting these special occasions.

Lighting for Entertainment

First, plan your lighting system for flexibility and creativity. Design it with extra outdoor receptacles on the circuits that serve the patio, deck, and other major outdoor areas, so that you can add temporary fixtures when you wish. Install dimmer switches instead of standard toggle switches to control most of the fixtures. This will enable you to focus just the kind of light and the intensity you want on each garden area for any occasion. Low-voltage lighting also provides flexibility. You can relocate an entire low-voltage system in a few hours by shifting the cable and fixtures, or install new fixtures by running cable from a transformer that has extra capacity.

Party lighting allows you to bring out the more spectacular side of your night garden. Use the lighting techniques that emphasize the small, subtle attractions of your planter areas, or the large-scale, dramatic focal points and vistas of the garden. You can create these lighting effects temporarily for a particular gathering by using dimmer switches to bring up the lighting level on fixtures that are usually kept dim, by plugging in temporary fixtures, or you can recapture a bit of lighting history by adding nonelectric light sources in special areas. Good lighting techniques to try are uplighting, mirror lighting, silhouetting, shadowing, and background

Above left: Paper bags, tiki torches, and colored balloons tell guests a party is awaiting them.
Above right: Mini lights make this dinner party a special occasion.

lighting that reflects light from plants or walls. Especially attractive are those types of accent lighting that bring the play of brilliants into the garden, with small temporary border and path lights, strings of mini-lights, and small accent lights of punched metal. See Chapter Two for more information on these techniques.

Suit the lighting mood to the occasion. A well-designed system can create the mood that's appropriate for any gathering. For a quiet gathering of close friends, light small garden areas intimately and provide a soft background of uplighted trees and shrubs. Add to this small accent lights focused on subtle garden textures and colors, and perhaps some mini-lights draped in a bed of ground cover or arrayed on an attractive shrub. For larger parties, you can bring out the spectacular, theatrical lighting effects that would be too overpowering for everyday use. Focus bright spotlights on the garden focal points of trees, shrubs, and statuary. Use perspective lighting to emphasize the garden's expanse and structure. Light garden water brightly as a reflecting pool, shimmering fountain, or glowing backdrop. But don't neglect to provide a few quiet havens of soft, diffuse light or near-darkness for those guests who want a retreat. You can create all these lighting effects temporarily through the use of dimmers and temporary fixtures.

One designer suggests that your permanent lighting should not focus too brightly on major outdoor sitting areas, so that you can add temporary lighting there with both electrical and non-electrical fixtures when you wish. This will create a special focus on the area.

Some temporary electric fixtures to use for party lighting:

- paper Japanese lanterns with small wattage incandescent lamps, powered by extension cords, for patio lighting
- temporary spotlights to uplight trees, shrubs, and house walls
- a "play of brilliants" using strings of mini-lights in ground cover or draped on shrubs and trees; also small accent lighting fixtures positioned in planter beds or at path corners, and punched copper fixtures.

Non-electric lighting fixtures for party lighting:

- candles in paper bags (fill bag partly with sand; you can also punch patterns in bag)
- votive candles surrounding a reflecting garden pool, or a stone lantern lit by a candle
- kerosene lamps on tables
- Hawaiian torches
- fire in a firepit.

Holiday Lighting

Your lighting system can also make an attractive display in your garden for holidays. Outdoor lights are frequently used to create special Christmas lighting; they can also brighten Halloween, Easter, and other special days. You can aim spotlights or accent lights at specimen plants, string clear mini-lights or Christmas lights on shrubs, trees, or the house, train small spotlights on decorative arrangements of dried materials near the front door, or uplight a group of carved or painted figures arranged on the lawn. You can also line the walk with candles in paper bags, or use other non-electric lighting for a festive atmosphere. (See the section above.) The guidelines that follow will help you create attractive and safe holiday lighting.

Work with one, simple theme. Don't plan to put one type or color of lighting in one area, another somewhere else. If you are planning a Christmas lighting display, for example, you might choose a theme of stars, or angels, or Santa Claus (but not all three), and decide to use mainly low voltage incandescent lighting to light it. If you are using colored lights, singly or in strings of Christmas lights, the display will be more dramatic if you limit them to one or two colors.

Focus the lighting in the most attractive part of your yard. Your holiday display will look best if it is a subtle, artful composition in one specific area, rather than being scattered all over the house and garden. A lighted display for Christmas or another holiday is usually intended to be seen from the street, so you may want to build it in the front yard. You can center it on a striking tree that you light attractively, or on a brightly lit, welcoming front porch. You may be accustomed to putting strings of colored lights all across the house frontage, but many homes look better with lighting that just outlines a balcony or window.

To light a Christmas nativity scene, aim small uplights set at ground level, from several directions. Alternatively, you can try background lighting of

plants or a nearby wall to make the figures stand out as silhouettes.

Trees are favorite subjects for holiday lighting. If you are stringing lights on a tree, whether Christmas lights or clear mini-lights, calculate how many lights you need like this: To make the tree solid with lights, figure the height times the largest diameter of the tree times 3 to get the number of bulbs you should use. To light with a special pattern or to accent the shape of the tree, multiply the height times the largest diameter times $1\frac{1}{2}$ to get the number of bulbs. You can string the lights in a random pattern, in a conical pattern by running vertical strings from the tip down to the base, in a spiral pattern from the top of the tree down, or in horizontal rows around the tree. Protect a living tree from damage by hanging the lighting wires from ornament hooks rather than from the small branches, and avoiding attaching a star to the growing tip (if you want a top ornament, strap a stick to the growing tip of the tree and attach the star to it).

Because strings of Christmas lights and other holiday fixtures may be used infrequently, check the cords for any sign of fraying or damaged insulation before using them, and make sure the lamps are working. If you are using many fixtures or strings of lights, try to spread the electrical load by plugging them into several different circuits.

Above: This attractive, yet simple lighting design brings the holiday spirit to the Dunsmuir House and Gardens in Oakland, California.

Left: This Christmas display uses colored lights to highlight the pine wreath and boughs.

LIGHTING AND PLANT GROWTH

Outdoor lighting can affect the growth of plants during the growing season, their flowering, and their dormancy in winter.

Plants living near bright artificial light sources may make more shoot and leaf growth as a result. These plants may reach their mature size more quickly; they may also require pruning more often.

Some annual and perennial flowers are called "long-day" or "short-day" plants, because they require one or the other to begin forming flower buds. Outdoor lighting can affect these as well, by extending the apparent length of the day. This will encourage early flowering in such long-day plants as carnations, China asters, and many annual flowers. It can delay the flowering of such short-day plants as poinsettias and chrysanthemums. You may have observed this without realizing it if you have ever kept a poinsettia indoors under artificial light for some months and then watched it wait until Easter to flower.

Some trees and shrubs growing in cold winter areas may not go properly dormant before the onset of cold weather if they are brightly lit by streetlights or other fixtures. The lights artificially extend the length of the day, and so confuse the plant's natural mechanism for sensing approaching cold weather. If a normally dormant tree or shrub is still actively growing when the cold temperatures do come, it can be damaged by the cold. Once it goes dormant, artificial lighting in early spring can cause early opening of leaf buds and potential damage from late frosts. In addition, some trees and shrubs require a certain length of dormant period to properly set flower or leaf buds for the next spring; if these plants are kept from going into dormancy by night lighting, they may leaf out irregularly or bloom less in the following spring. Some plants especially sensitive to these types of damage are London plane tree (*Platanus x acerifolia*) and birches (*Betula* species).

OUTDOOR LIGHTING AND INSECTS

There's no question about it: outdoor lighting can draw insects that make outdoor living miserable. Here are some ways of dealing with this problem.

1. Locate bright floodlights and spotlights needed for games and entertainment areas well away from the center of the area. These bright lights are the worst bug attractants. You can mount them high overhead, as you would, for example, if you were moonlighting, or at some distance across the yard and aimed back toward the area. This way the insects will be attracted to the light sources, and less to you.

2. Use different types of lamps to selectively attract insects. Most insects are more attracted to blue light than to white or yellow, so you can use several mercury vapor lamps in the more distant garden, and standard incandescent lamps near where people congregate. One designer uses yellow-tinted PAR incandescent lamps (these are commonly available) to contrast even more markedly with the blue mercury vapor lamps.

3. Use light only when and where you need it. This not only helps the energy efficiency of the system (see page 31); using less light will draw fewer insects. If you only turn on light where you need it and use dimmers to provide just the amount of light you need, fewer bugs will find you.

4. Buy commercially available "bug-lights." These lamps are specially coated to exclude blue light. Because insects are attracted especially by the blue content in most light, they do not "see" the light from these yellow lamps. These lamps do have several drawbacks, though. They are not very energy-efficient, having a particularly low lumens-to-watt ratio (see page 93), and the color rendering is not attractive, making most green plants appear sick or dead, and complexions look unnatural.

5. If none of these other measures work for you, consider installing an insect trap. These traps use various attractants to draw insects and then kill them. They can effectively limit the number of insects plaguing you.

Yellow "bug lights" attract fewer insects than standard lamps.

THE DESIGN PROCESS

The design process includes assessing your property and lighting needs, experimenting with lighting techniques and fixture placement, calculating your electrical power needs, choosing materials, and drawing the lighting plan. The process is logical, and you'll find that each step will bring you closer to knowing what you want to light in your garden and how to do it.

STEP 1: *Why Are You Lighting?* First, look at how you use your garden and what lighting you need to make it more attractive and safe. Think about and observe your own habits, the garden, and your existing lighting systems both in daytime and at night. The questionnaire below will help you do this.

Favorite Garden Features

- What do you like best about your garden?
- What individual features of the garden do you most enjoy? These might be favorite plants, attractive pavings, architectural elements like fences, masonry walls, or a wood deck. Your lighting scheme can enhance these features.
- What are your favorite views of the garden from inside and outside? Include here views of the garden or views that go beyond the garden to a distant range of hills or a specimen tree in your neighbor's yard.
- How do these views change with the season?
- Where do you sit to enjoy these views?

Least Favorite Garden Features

- What bothers you most about your garden?
- What do you wish you didn't have to look at out the window? Include here visual nuisances both in your own yard and those that lie beyond it, such as your garbage cans, a neighbor's house, or power lines along the street. At least at night, you can control what is visible by lighting only what you want to see.

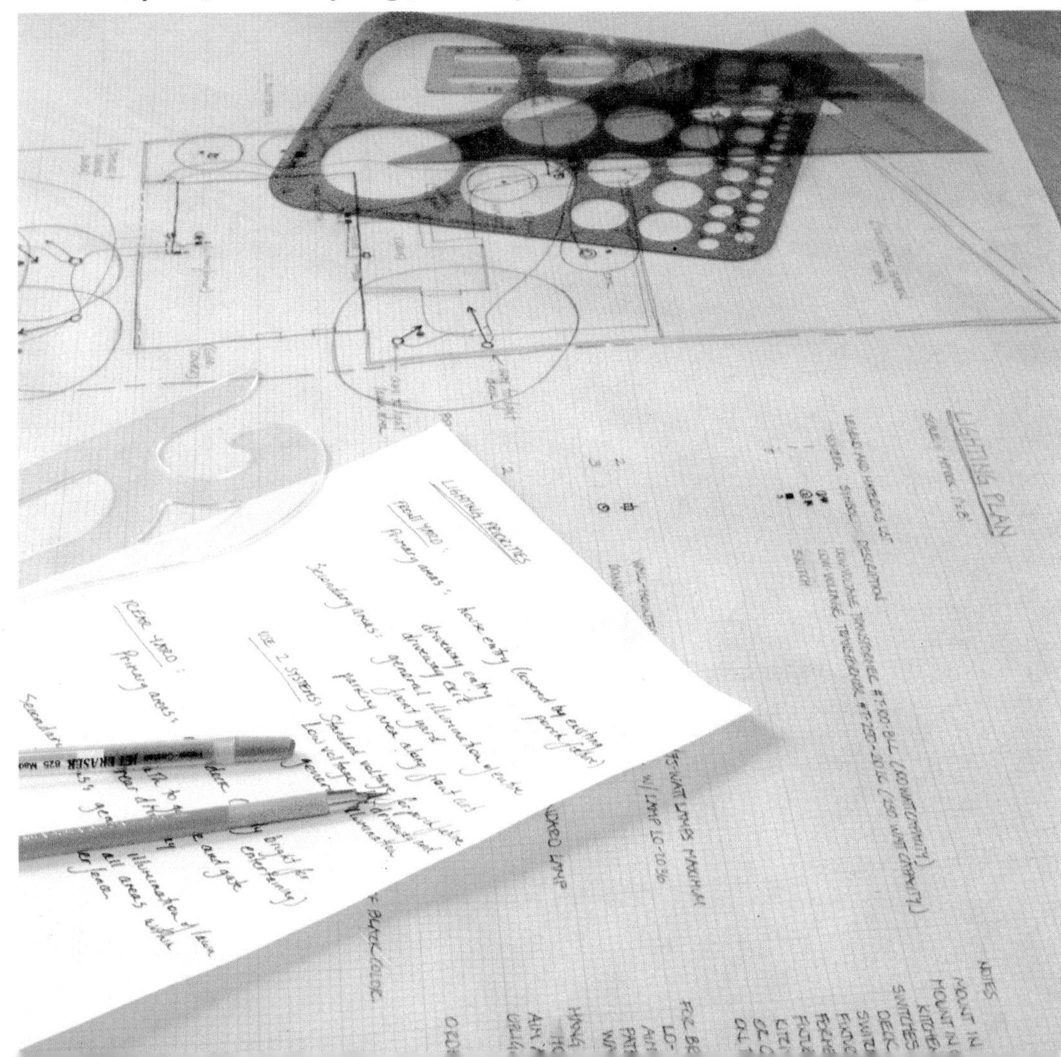

A plan helps you organize your thoughts and ideas about the lighting design.

This formal garden has a strong central axis.

Design Concepts

■ How did the architect, landscape architect, and interior designer conceive of your home and grounds?

■ What good points of your site was each working to bring out, and what negative features did they try to overcome? Even if you did all your own design, try to remember the goals you had in mind and plan to strengthen them with your lighting.

■ What are the axes, or main sight lines, of the garden? These might be the view from one end of the yard to the other across a patio, the open space between rows of trees where a path winds, or a long, narrow lawn.

■ What are the main spaces, or "outdoor rooms" of your garden? Are they enclosed by walls of masonry or plantings, or beneath the ceiling of an arbor or tree canopy?

■ What plants and materials form the "floor" of your landscape?

■ What form, color, and texture do the individual plants and materials have?

■ Where are the "edges," where one plant type or material meets another?

■ What colors, shapes, and textures have you used in the rooms indoors?

■ What about the house exterior? Is it the rich brown or weathered gray of wood siding? If it is painted, how light and reflective is the color?

■ Is the style of the house angular and modern or more traditional?

As you answer all these questions, you can begin to sense the design themes of your home and garden: how different areas fit together, their important and attractive elements, and what is irrelevant or ugly in the garden. You can design your garden lighting to unify the design themes of interior and landscape, conceal what is unattractive, and shape the night view of the garden.

How You Use Your Garden

■ How do you use your garden, in daytime and at night?

■ Do you more often view it from indoors or use it for outdoor activities?

■ Where in the garden do you do your relaxing, entertaining, dining, strolling, outdoor work projects, or play active games?

■ What other garden activities do you enjoy?

■ Where would you do these things at night if the areas were lighted?

■ How much light does each activity need?

The Moods of the Garden

■ How would you like different areas to feel? Do you want the patio to feel relaxed, the entry to be inviting, the spa to have a romantic glow?

Lighting can help create all these moods, depending on how you place the fixtures, how bright they are, and even the color of light used.

What About Visitors?

■ Where do visitors normally enter the property?

■ What route do they take to the front door or garden?

■ What activities will they enjoy in the garden? Where?

Garden lighting can illuminate their path, light their activities, influence their moods, and help them feel comfortable in the garden.

Safety

■ What parts of the garden are dangerous at night? Include steps, other changes of grade such as retaining walls, water elements, uneven paving on paths.

■ Which of these are most dangerous, or get the most use?

Security

■ How secure are your house and ground against burglars and other intruders? Consider the neighborhood, local crime rates, the presence of streetlights, whether there are dark shadows near the garage, front porch, or windows.

■ Where might intruders enter the property? These areas can be the focus of security lighting.

■ How secure do you feel about your home? Good security lighting can make you feel safer and more comfortable in your house and garden.

Existing Lighting

■ What does your existing interior lighting look like viewed from outside at night?

■ Is your existing outdoor lighting attractive?

■ Do any nearby streetlights illuminate your property?

This lighting system was planned for nighttime activities such as swimming, dining, and entertaining.

- Does the existing outdoor lighting blend with the interior lighting?
- How effectively does it light for beauty or effect, activity, safety and security?
- How could it be improved?

Your answers to these questions should start you thinking about where you need lighting in your garden and how much to light each area. When you're considering how to improve lighting for beauty, safety and security, decide which garden areas are most important and which are less essential. Write down a list of priority areas for brightest lighting, secondary areas for background or fill lighting, and areas which should remain dim or dark to hide ugliness or provide contrast to the brighter areas. This information can now be mapped on a plan of your property.

STEP 2: *Making a Base Plan of Your Property*

A landscape plan is a convenient way of representing your garden on paper. It is a drawing of the yard as though you were looking down on it from high above, so that buildings, for instance, are drawn as simple geometric shapes, and trees as flat circles. The base plan for your lighting system is a summary of the most important elements of your existing garden and outdoor lighting. In later steps of the design process, you will add information about what you want to light, where the fixtures should be, and other materials needed.

If you are putting in only two or three standard-voltage fixtures or a single low-voltage system, you need not draw the plan to scale, but can simply sketch the yard as accurately as possible, noting on the sketch the measurements between major landscape elements. If your system will be more complex, you should draw the plan to scale.

The most common scale for landscape plans is 1 inch equals 8 feet, called $\frac{1}{8}$ scale. In this scale, one inch on your plan represents 8 feet of measured distance in the garden. Other scales used are 1 inch equals 4 feet for small gardens, or 1 inch equals 16 feet for large areas.

To make your plan, first gather the following drafting supplies:

- several sheets of graph paper in the scale you will work in (for example, paper with 8 squares per inch for $\frac{1}{8}$ scale). The best size is 24″ by 36″.
- a roll of inexpensive tracing paper, or loose sheets, in the same size.
- a straight edge, or for more professional results, an architect's scale. This enables you to easily read the inch equivalents of your measurements in the garden.
- several soft-lead drafting pencils, in grades F or B.
- a roll of masking or drafting tape.
- a smooth table or drafting board, large enough for a full sheet of graph paper.
- a 50′ or 100′ tape for measuring your garden.

If your lot is large or irregular in shape, the easiest way to begin mapping it is to get a copy of the assessor's parcel map for the property. This shows the dimensions of your lot, and so saves you some measuring. The parcel map will often be attached to the title report that you got when buying the house, or can be obtained from the county assessor's office. If your lot is small and rectangular, or you can't get a parcel map, simply measure your property lines with the tape, noting the angles where they meet at the property corners. Draw the outline of your property in the center of your paper, to scale, with north at the left or top of the plan.

Now find a *base line*, which is a straight line from which you can measure distances to objects in the landscape. A good base line is often one outside wall of your house, or a well-marked property line. Measure from the base line to the important objects in the landscape. These include other house walls, doors and windows, structures such as an arbor, garage, gazebo or pool house, paved areas and paths, pools, the trunks and leaf canopies of trees, locations of major shrubs, and the outlines of other planter areas. Record also approximate locations of electrical service panels, outdoor electric receptacles, and existing outdoor lighting fixtures.

Translate these measurements to the scaled plan or sketch, and map them in pencil. This is your base plan. Take good care of it: even after you've used it to design your lighting, it may be useful for planning other landscape work or remodeling.

Steps and other hazardous areas should be brightly lit.

STEP 3: *The Site Plan*

Lay a clean sheet of tracing paper over your completed base plan. Record on the tracing paper the information and ideas you generated in Step 1, above. When you are finished, it should indicate the major garden areas and axes, the paths and other connections between them, views and focal points, outdoor activities, visitors' points of entry and circulation routes, safety hazards, areas to light for security, features to accent or conceal, and garden moods to create. Note comments on existing lighting fixtures and make preliminary notes on new lighting.

STEP 4: *Establishing Lighting Priorities*

Now lay this site plan aside, and put a fresh page of tracing paper over your base plan. Draw a "bubble diagram" of your lighting ideas, by sketching on the tracing paper loose shapes, or "bubbles" representing important garden elements to light, areas for brightest lighting, areas for medium-bright (secondary) lighting, and those to be left dark or dim. Refer back to the site plan and decide what garden elements and areas are most important to light, which are of secondary importance, and what areas are unimportant, irrelevant, or should be left dark for contrast. For help in deciding how bright each area should be, you can also reread the section on levels of brightness, on page 8. Consider not only safety and security needs but also what should be lit for beauty, to emphasize design themes or create a mood, and for garden activities, entertaining, and games.

Primary areas for brightest lighting will include those garden features that will be focal points of your lighting scheme. These may include a large oak tree that graces your backyard, or a statue or wood arbor that plays a major role in the landscape design. Other areas for bright lighting are active game areas; garden work centers; important safety hazards such as steps, water edges, and major changes of level in the garden; heavily used circulation routes such as the entrance to the driveway from the street, and the walks from the street and garage to the front door. You may also wish to light brightly the far end of a garden axis for perspective lighting (see page 25), or some parts of the property for security.

Secondary lighting connects the bright focal points of the garden by lighting less brightly the circulation routes, the "middle ground" along garden axes for perspective, and secondary safety hazards. It also provides the low-level general illumination used in moonlighting and large-area security lighting.

Also plan to leave some areas dark, or lit only by a dim spill of light from distant focal points. This gives the night garden contrast, hides what you don't want to see, and creates an air of serenity and mystery.

Draw several different bubble diagrams of your garden to try out different ideas. Note on the diagrams the emotional feeling you want each area to have, whether this is relaxed, stimulating, or mysterious. Look back to the discussion of lighting techniques in Chapter Two, and make notes of techniques to try in different areas. For additional guidance and specific suggestions on lighting techniques, fixtures, and lamps for each part of the garden, refer to the section on Ideas for Lighting, page 49. Spend some time in the garden at night trying to visualize your ideas and then modify the diagrams as you get a feeling for what you want. Finally, narrow down your ideas to one or two diagrams, and retrace them if they're cluttered or messy.

STEP 5: *Trying It Out*

Now you can start bringing your lighting ideas to life. There is no substitute for direct experimentation with different lighting techniques at night in the garden. Your guide for this testing is the bubble diagram you just completed, which suggests lighting intensities, techniques, and mood for each outdoor area.

To begin, gather a few fixtures and supplies. Get one or two 100′ heavy-duty extension cords, a plastic "trouble light" of the type used by mechanics (this and the extension cords can often be rented), a reflector-fitted portable lamp with a grounded plug, Type A incandescent lamps of 15, 25, 40, 60, 75 and 100 watts, PAR incandescent floodlamps of 75 and 150 watts, and PAR spotlights of the same wattages. Also get a number of short wooden stakes for marking fixture locations and an indelible marking pen for making notes on the stakes. If you are planning to use mainly low-voltage lighting, try to borrow a low-voltage demonstration kit from your supplier, or purchase a transformer and a few fixtures: one for spotlighting, a mushroom or pagoda-type fixture for path lighting, and one fixture for large-area lighting or downlighting. You are likely to use these fixtures in your final scheme, and you can buy the rest of the low-voltage fixtures you need when you've finished the design.

While trying out your lighting ideas, keep safety in mind. Do any connections of fixture and lamp, or fixture and extension cord, with the power off. If you are joining several extension cords for distant fixtures, tape the joints between them with electrician's tape. And don't work when the ground is wet.

Begin to try out lighting techniques at night. Plan to locate only a few fixtures in a night of work, having the patience to try many different locations and angles. The trouble light can imitate an uplight for silhouetting, shadowing, or accent lighting; you may wish to make a shield of aluminum foil covering part of the fixture on the inside so you can see the lighting effects and not just the fixture's glare. The reflector lamp can take either PAR or A lamps, and can be used to suggest all kinds of downlighting effects, grazing light, spotlighting (use the spotlight), or mirror lighting. Try different lamp sizes according to the brightness needed in each spot. Don't use any of these fixtures underwater. Drive a stake in the ground at the location of each fixture, and write on it notes about fixture type and position, and lamp size.

Decide first how to light the largest trees in the yard, or where to put other high fixtures for downlighting. Light from these fixtures can provide much of the illumination you want for beauty, safety, and security below. (See the notes on lighting of trees on page 56 in this chapter.) Downlighting fixtures can be hung high on house walls or other buildings as well as in trees, but trees provide a more natural shielding of the lamp. Work outward from the trees and other focal points in the garden to the secondary areas, noting how much light is already provided by spill from the brightest areas and where to add fixtures. Don't make the pattern of fixtures too uniform, even along paths, but concentrate the light where you need it. Plan also where to put outdoor electrical receptacles. These will come in especially handy around entertainment and game areas, but you may choose to design nearly all your lighting as "temporary" fixtures plugged into receptacles.

Work for unity and perspective in the garden by lighting foreground viewing areas brightly, the "middle ground" beyond this dimly so that you can see shadows and silhouettes, and the background or end of the view more brightly with highlighted trees, backlit shrubs, and so on. See the notes on perspective lighting in Chapter Two, page 25.

When you've decided how to light the objects and areas of your garden, draw in the fixture locations on a tracing paper overlay of your base plan. Also note on this sketch the lighting techniques to be used, the lamp types and sizes, and any other ideas about positioning the fixtures.

Take a little time now to walk in the garden, visualizing the new lighting you've planned. What will be the feeling of each area when lighted? How far will the light spread from focal points and main activity areas? How will secondary lighting connect these brighter areas? What will be your experience as you walk along paths, up to the front door from the garage, or sit on the deck? Do all the different lighting techniques and effects fit together into a harmonious whole, making the garden more beautiful and useful? Make any changes now that will improve the system, and revise your overlay sketch accordingly.

If you are having a contractor install your system, you can stop here, trace your ideas onto your base plan, and show the plan to the contractor as a summary of your lighting needs. Or you can go on to complete Step 6, so that you can decide which areas should be served by each outdoor lighting circuit and which fixtures each switch should control. The contractor can check your circuit layout and electrical loads, and proceed to install the system.

AN ALTERNATE DESIGN METHOD: A SYSTEM OF "TEMPORARY" FIXTURES.

One landscape architect, James van Sweden of Washington, D.C., designs lighting systems for his clients' gardens by a very simple method. He lays out a network of double outdoor receptacles every 15 or 20 feet throughout the garden, then he provides the client with a number of basic standard-voltage fixtures for downlighting, uplighting, safety lighting, and area lighting, which can then be used wherever the client wishes as "temporary" fixtures. They can be moved seasonally to focus on a flowering tree, a verdant lawn, or newly fallen snow. Van Sweden may install a few permanent fixtures along major paths and at the front entry, but otherwise the system is designed for complete flexibility.

Use "temporary" or movable fixtures to call attention to seasonal color or blossoms.

STEP 6: *Calculating Power Needs*

Now you need to determine whether your existing electrical circuitry can handle the extra load of your new lighting, or if you must add new circuits. This calculation is an important one: if you add your lighting to an existing circuit already being used to full capacity, you will overload the circuit, and you can expect it to continually blow fuses, trip circuit breakers, or cause more serious problems. How to calculate your circuit wattage and lighting demands is explained in detail below, but if you are at all uncomfortable making the calculation yourself, consult an electrician.

First, map the existing circuit you are considering using by determining all receptacles, switches, and light fixtures that are supplied by that circuit. You can do this by turning on all indoor and outdoor fixtures and small appliances, and plugging additional lamps or small appliances into all unused receptacles. Then unscrew the fuse or turn the circuit breaker to "off." Note what lights and appliances go off and put this information on a quick floor plan of the house, or simply list the rooms and outdoor fixtures served by the circuit. Receptacles that do not have an assigned use are usually allotted 180 watts (1.5 amps) for load calculations. Add up the total watts of electricity consumed by the lamps and small appliances normally on the circuit. You may want to map several circuits to find the one with the greatest unused capacity.

Now determine the capacity of the circuit, which is how many watts of electricity it can safely carry. If the fuse or circuit breaker is rated for 15 amps, you should use a maximum of 1500 watts on the circuit. For a 20-amp circuit, use a maximum of 1800 watts. Do not use more than a 20-amp circuit for outdoor lighting.

Next, subtract the present wattage load on the circuit from the capacity of that circuit to see if there is unused capacity. For example, suppose you find by mapping a circuit that it currently uses approximately 1000 watts when all receptacles and fixtures are being used. This circuit is fused by a 15-amp fuse, so its total usable capacity is 1500 watts. This leaves a surplus of 500 watts for outdoor lighting.

Now figure the total wattage of the new outdoor lighting you are planning. This can be an approximate figure, based on the lamps you expect to use in the outdoor fixtures and the additional temporary lamps or machinery you may plug into the new outdoor receptacles. If this total is less than the unused capacity of the existing circuit, you may be able to add the new lighting to that circuit. If you are putting in low-voltage lighting, the total wattage required may be just several hundred watts, and you can safely add this to almost any existing circuit. But if your new lighting is standard-voltage, be more cautious. Don't plan to add more than two standard-voltage fixtures to an existing circuit that includes any receptacles; any additional appliances or fixtures plugged into the receptacles later may cause an unexpected overload.

Take into account also two other factors: the wire size to be used on the new lighting and the length of the planned wire runs. The length of wire runs on electrical circuits must be limited to avoid "voltage drop." This is the loss of voltage on a circuit due to excessive length of wire or an excessive amount of electrical current, and is usually seen on lighting systems as a dimming of the lamps. The greater the amount of electricity on a circuit, the shorter the distance it can be carried and the larger the wire or cable needed to carry it. The table gives suggested maximum wire run lengths for the three commonly used sizes of wire on 120-volt lighting circuits using different total amounts of electricity. The table is based on recommendations of the National Electrical Code and is intended as a guide only; you can install circuits with wire runs longer than it recommends provided that you wire them correctly and don't mind dimmer light at the end of the line due to voltage drop.

To use the table, calculate the approximate total watts of electricity you'll use in your lighting system, and look up the recommended wire sizes and wire run length in the opposite columns. Be sure the circuit is protected by a fuse or circuit breaker of the required capacity.

When figuring wire length, measure both the length of the existing indoor circuit and that of the new outdoor wiring. To measure the indoor circuit, measure the approximate distance from the service panel to the point where you plan to connect to the circuit and add 50 percent to this measurement for safety. Then combine this figure with the estimated length of outdoor wire runs.

Maximum Length of Wire Run on Lighting Circuits[1]

Total Watts[2]	Maximum Wire Run (in feet)[3]			Fuse or Breaker Rating (in amps)
	#14 wire	#12 wire	#10 wire	
100	820	1370	2190	15 or 20
200	410	685	1095	15 or 20
500	165	275	440	15 or 20
800	100	170	270	15 or 20
1200	70	115	185	15 or 20
1500	55	90	145	15 or 20
1800	45	75	120	20

[1]This table applies only to 120-volt lighting systems. For 12-volt systems, follow the manufacturer's recommendations regarding wire size and length of wire runs.

[2]The maximum wire runs are assuming the total load is at the end of the circuit. When the load is distributed over the length of the circuit, the total length can be greatly increased before noticeable dimming of lamps occurs.

[3]Length of wire run refers to distance from the circuit box to the light fixture at the end of the circuit.

Top: Multiple fixtures and long runs of wiring may require more than one circuit.
Bottom: Lights that will be controlled as a single unit can be placed on the same circuit.

As an example, if your existing indoor load on a circuit is 1000 watts and your proposed outdoor lighting would add to it 500 watts, the total load on the circuit would then be 1500 watts. If the circuit is fused for 20 amps, its capacity is 1800 watts, so you can safely add the outdoor lighting to the circuit if you use #12 cable and the total distance to the end of the new combined circuit is 90 feet or less. Alternatively, if the existing circuit uses #10 wire, you could add additional #10 wire to it for a maximum combined wire run of 145 feet. To take another example, if you have a 15-amp circuit with an existing load for indoor uses of 1200 watts and your outdoor lighting would add 500 watts, this would bring the total wattage on the circuit to 1700 watts, 200 watts more than the maximum of 1500 watts. You cannot add your lighting to this circuit, no matter what the wire size or wire run length. Look for another circuit with greater unused capacity, or consider adding a new circuit.

Now make a map of your new outdoor lighting fixtures on a tracing-paper overlay of your base plan, and divide them into circuits. To do this, first mark the locations of the fixtures on the plan and the approximate watts of electricity each will use. If there are more than a few fixtures you will probably want to divide them into separate circuits according to the different garden areas each serves and the wattage required. You can also combine different areas on the same electrical circuit, operating them by separate switches. The calculations and table above will tell you if you can safely add some of the new lighting to existing circuits, or if you must wire it as new circuits from the service panel. Indicate on the plan overlay each circuit and its point of connection to power, and the fixtures on each circuit operated by each switch.

If you have determined that you must install new circuits for your lighting, you can also use the table above to determine the wire size, required fuse or circuit breaker, and maximum wire runs on the new circuits. For example, if you lay out your system on paper and find that on one circuit the length of wire from the service panel to the farthest fixture will be 400 feet and the circuit will use 200 watts, you see from the table that you can safely use either #14 or #12 wire. But if you are thinking of adding another 400 feet of cable to this later for additional fixtures, the total combined wire run length will be 800 feet, and you should use #10 wire for the entire circuit.

If you are planning a standard-voltage lighting system, you may have made the calculations above and decided you need to add one or two new outdoor lighting circuits. If your system will be quite complex, you may even have decided to install a second electrical service panel to operate it. Adding circuits or a new service panel can be expensive; these are also exacting jobs for which you may want to call in a professional electrician.

STEP 7: *Choosing Materials*

Now the hardest design work is behind you. You have already made some guesses about the materials you'll use for lighting, in order to do the calculations in Step 6; now you can choose the specific fixtures, lamps, and other materials that will make up your system. For a complete survey of what is available, refer back to Chapter Two.

Think again about the advantages of low-voltage lighting systems, particularly if your calculation of the power required to run a standard-voltage system suggests that you need new circuits or another service panel. Low-voltage lighting may be able to save you much trouble and expense.

Look again at the descriptions of different lamps in Chapter Two. Will you choose incandescent or mercury vapor, or combine these with fluorescent, high-pressure sodium, or metal halide?

What about fixtures? Get catalogs from your lighting supplier or hardware, or write directly to the manufacturers listed on page 39. In deciding which product lines and individual fixtures can best meet your needs, consider several factors:

- What function does the fixture need to serve? Should it be a downlighting fixture with an extra long shield to hide the lamp? A porch lantern that must be equipped with a diffusing lens because it will hang near eye level? A watertight underwater spotlight for a fountain? How large an area does each fixture need to light? Fixtures are available to meet many very specific needs.
- What size fixture is most appropriate for each spot? You may need large floodlighting fixtures for a grand expanse of lawn, or a small accent spotlight for an apartment deck planter. Choose a line of fixtures that is appropriately sized for your garden.
- What about the design of the fixtures? Should they be especially attractive for prominent places on the deck or patio, or unobtrusive for easy concealment in trees and shrubbery? What entry fixtures will best match the house interior and exterior? Can you get all the different fixtures you need from one manufacturer so they all have a similar appearance? What additional design requirements should they meet?
- How flexible are the fixtures? Can an uplighting fixture also be hung in a tree as a downlight? Can you adjust the aiming angle of the fixture or the lamp within? If your system uses standard voltage, some manufacturers make a stake-and-cord mounting assembly that enables you to reposition the fixture slightly. And if you want maximum flexibility, choose low-voltage lighting; fixtures and cable on these systems can be simply dug up and moved.
- The cost and durability of the fixtures are also important. As with most products, you can get more durability if you spend more. Some lighting manufacturers make fixtures of wonderful quality, which are unfortunately beyond most homeowners' budgets. Other product lines are very inexpensive, but the fixtures may be built of lightweight plastic, or with poor quality electrical sockets, and will come apart within a couple of years. Ask your supplier which products offer the best value within the range of your budget and check any guarantees that come with them.

Also choose at this time the other materials that will make up your system: cable, conduit, boxes, switches, dimmers, time clock, and low-voltage transformer. To determine the size

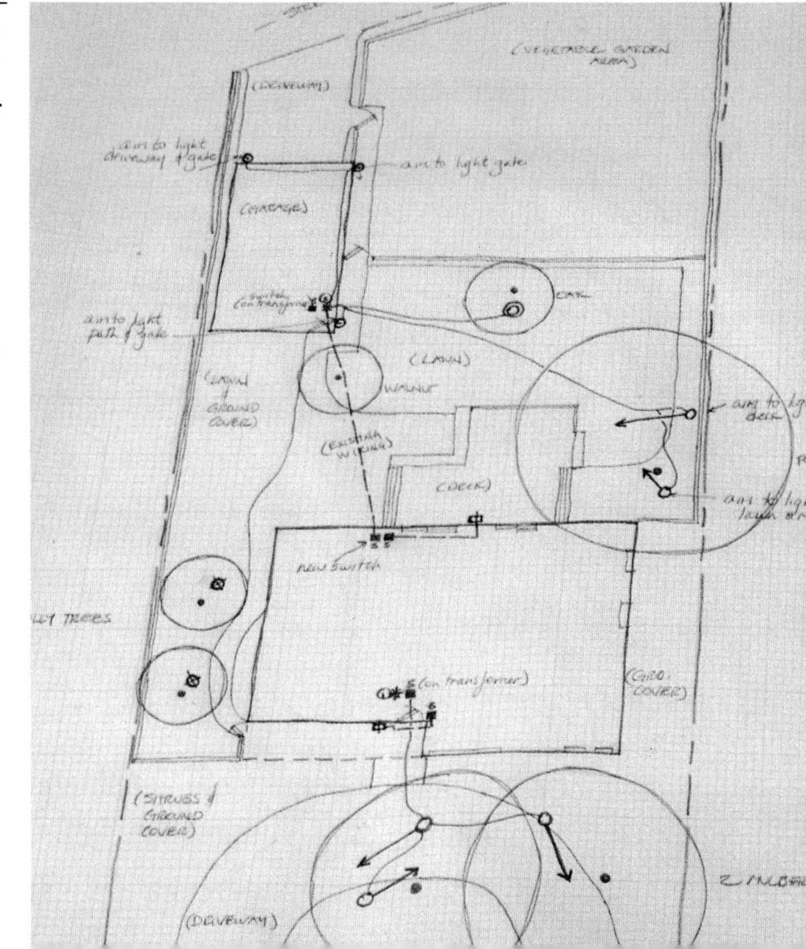

The completed lighting plan will be used by you or your contractor to install your lighting system.

transformer needed for a low-voltage system, simply add up the watts of all the lamps you expect to use on the system and order a transformer with capacity to spare.

Look for most of these materials at your lighting supplier or local hardware store. How much you spend here, too, can determine the quality you get. Be certain all materials are approved for use outdoors and buy the best quality you can afford.

After you have chosen the fixtures and lamps, return to your calculations and the circuit map you did in Step 6. Revise them and remap the circuits according to your new materials list.

STEP 8: *The Final Plan*

Now you've gathered all the information you need to draw the final lighting plan. You've chosen the lighting techniques to use, you know which fixtures, lamps, and other materials will work best for you and which fixtures will be switched together. You've divided all fixtures and receptacles into circuits and determined where each circuit will be connected to power.

Now return to your base plan, drawn in Step 2 above. Choose symbols for each type of fixture you'll use, so they can be easily represented on the plan. If a particular fixture will use different lamps in different locations, designate each fixture-lamp combination with a separate symbol. Connect them into switching patterns and circuits; the lines connecting the fixtures will represent the approximate layout of the wire runs. Indicate the location of the main service panel and the point of connection of each circuit to power, whether this is to an existing circuit or to the panel. Use a legend in a corner of your sheet of graph paper to explain fixture symbols with manufacturers' names and model numbers. Also make notes wherever they fit on the drawing about the specific position and aiming angle of each fixture, different wire sizes where used, and other materials needed for electrical connections or the hookup to the power source.

If your base plan is a sketch, not drawn to scale, indicate all critical measurements on the plan; these include the distances between major trees, across the patio, or between other important landscape elements, and the length of the wire runs.

Calculate the cost of your system by adding up the cost of all lamps, fixtures, cable, conduit, time clocks and other materials; the fees to be paid for building permits if required; and the cost of any work you will have done by a contractor. This last may include a consultation fee for having the plan checked (see below), the cost of having the electrician inspect your final installation (a step we recommend if the system won't be inspected by the local electrical inspector), and the charges for any other part of the installation you feel safer leaving to a professional.

STEP 9: *Getting It Checked*

If you are planning to install your own lighting system, you should now take your completed plan to a licensed electrician to have it checked. Most electricians will do this for a consulting fee and can suggest corrections you should make to the design.

Once you have had the final plan checked, your lighting design is complete. Now you can decide how much of the system installation to do yourself and how much to have done by professionals. Full details of the installation process are given in Chapter Four; if you feel comfortable with electrical work, or are installing a simple low-voltage system, do the work yourself. Or you can hire an electrician to install part or all of it for you. In either case, the electrician or local inspector should check any parts of the system installation you do.

If your budget is limited, your lighting system can also be installed in stages, putting in first the major wire runs from the power source to the areas that will be lighted (without connecting the wiring to power), and adding the fixtures in each area later as you are able. If you install your system piecemeal in this way, plan to finish the lighting for your main patio or outdoor sitting area first because this will be the outdoor area you'll use most at night. Next, finish the lighting from the front parking area or garage to the front door, for safety and security. Third, provide attractive lighting in areas where guests enter the property or congregate outdoors, and finally, add aesthetic lighting to beautify the entire garden.

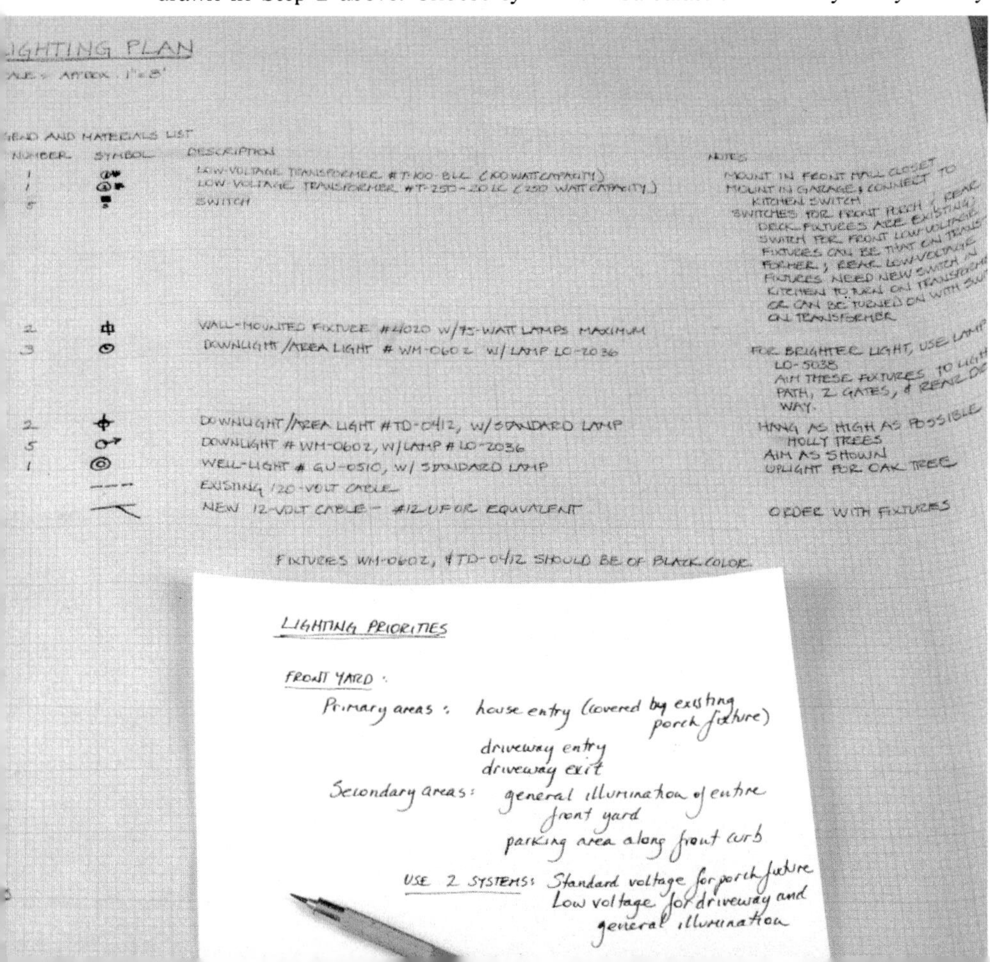

Installation and Maintenance

Step-by-step procedures to install your own lighting

Now that you've designed your lighting system, you are probably eager to see it become a reality. The hardest part of the job—that of working out your lighting needs and wants and setting them down on paper in a lighting plan—is behind you. All that remains is to follow the installation instructions in this chapter.

Each task you'll need to complete to install a 120 or a 12-volt system is clearly explained in the step-by-step instructions in this chapter. Before you know it, you'll be laying out the materials, connecting wire to the fixtures, and lighting your garden or deck. Once you've finished the job, it will be time to see how well your ideas on paper work with real trees, shrubs, and walkways. A flip of the switch and suddenly the branches of that oak tree come to life, the leaves of a fern seem to glow, and your patio becomes a nighttime entertainment center.

You needn't be a skilled electrician to put in your own lighting. If you obey the basic safety rules for working with electricity and do the work carefully, you can install a complete, successful lighting system yourself. Once your lighting system is in, you can keep it in the best condition with routine maintenance tasks.

A well-designed lighting system should provide you with years of enjoyment. The compliments you receive will be especially gratifying for you—the designer and installer who created all this beauty.

Opposite: You needn't be a skilled electrician to install a professional looking lighting system.
Right: The installation process consists of a series of easy-to-follow steps. Follow these steps, paying close attention to the instructions and you will soon be putting the final touches on your outdoor lighting system.

INSTALLING 120-VOLT LIGHTING

The main differences between 120-volt and 12-volt lighting installations are in the depth of burial of the cable, the permits needed for the work, the requirements for conduit, and the complexity of the wiring. A standard-voltage system is usually more expensive than a low-voltage system. The materials cost more, and you are more likely to have an electrician do part of the work. However, standard-voltage systems light your garden more brightly and may last much longer.

If your 120-volt outdoor lighting is being installed as part of house remodeling or landscaping, plan carefully when to do each part of the installation to minimize your work. You can do indoor wiring to switches easily while the wallboard is torn out, and quick conduit installation outdoors after rototilling of the lawn. See also the notes on page 78.

STEP 1: *Get Any Required Permits*

Begin by getting a building permit, if it is required for outdoor wiring projects in your area. This will probably involve taking your final plan to the building inspection office for approval. You may be asked to make changes in the plan to make it conform to the local electrical code, or to demonstrate that you are proficient enough to do your own work. The inspector will tell you at which points during the installation the work should be inspected.

Local hardware and lighting stores carry many of the materials you will need for your lighting system.

STEP 2: *Purchase Materials*

From your final plan, list the parts you need to build your lighting system. List the materials systematically, beginning with the lighting fixtures, conduit, conduit fittings, boxes, receptacles, switches, and cable. Don't forget wire connectors, GFI's, and dimmers. Refer back to Chapter Two for a discussion of these and other parts, and see the comments below.

Conduit

Decide now whether you'll use rigid PVC or rigid metal conduit. Your code may already make this choice for you, but if it doesn't, remember that rigid metal conduit is harder to work with and may corrode in time, but that it is also more resistant to accidental damage and rodents and can be buried more shallowly in most areas. PVC conduit is lightweight, easy to work with, and will not corrode, but may be chewed by rodents and in most locations must be buried at least 18 inches deep. Measure on your plan the length of conduit needed, add several feet to cover the vertical sections adjacent to fixtures, switches, receptacles, and wire runs into trees (if the code requires conduit here), then add 10 percent as a margin for error.

Boxes

You'll need one outdoor box for each fixture, receptacle, and switch. The only exceptions are the boxes for those switches that will be located indoors: these can be standard interior boxes. If you plan to control any lights from two separate switches, remember to order extra boxes for this. For wet locations, or to place the boxes at ground level beneath fixtures, you can order special junction boxes made of cast iron.

Standard outdoor boxes of the same type can be used for outdoor switches, junction boxes, and receptacles. You simply order a different cover for each location, according to its use. List on your parts list the number of each type of cover for the boxes, whether receptacles, switches, or solid covers for junction boxes. Standard outdoor receptacles have spring-loaded doors that protect the socket openings from moisture. Switches may be single-pole to control the lights from just one location or three-way switches to control fixtures from two points. Most switches should be located centrally indoors, and for the greatest lighting flexibility and energy efficiency, should be dimmers. Order dimmers that have a capacity (in watts) 25 to 30 percent greater than the wattage on the circuit; they'll last longer.

Wire

What kind of wire you order depends on what your local code permits. Type UF cable can be directly buried in most areas, so it is easier to work with. Type TW or THHN wire must be enclosed in conduit, and you need to buy individual wires to combine as the hot, neutral, and grounding wires in the installation. Measure the lengths of the wire runs on your plan, from the power source to the end of each run. As with conduit, add several feet for each above-ground connection, then add 10 percent for a margin of error. Get 2-wire UF cable with a ground wire for most uses, or 3-wire cable with ground for wiring three-way switches (see page 40).

Fittings

To join the conduit, boxes, and fixtures together, you'll need couplings (slip-type couplings for PVC conduit, threaded metal or compression couplings for metal conduit), and threaded PVC fittings for attaching PVC conduit to the boxes. Do not use any set-screw couplings to join metal conduit outdoors as they are not watertight. Order some insulated bushings to put on the end of metal conduit to prevent damage to the insulation when pulling cable through. You'll also need sweep bends for making 90-degree bends in PVC conduit; do not use the smaller bent fittings sold for PVC water pipe.

Wire Connectors

The best connector for outdoor wiring is the wire nut. Be sure to buy the right size for the wire size you're using, so that it will fit snugly.

Time Clock

See the discussion of timers on page 42, and choose the type that meets your needs. For security lighting choose one that varies the hours of operation slightly each night.

Other Switches and Sensors

For easy control of your outdoor lighting from anywhere indoors, you can order plug-in switches. These will operate any receptacle to which you've attached a plugged-in or wired-in module. See page 42 for a discussion of other types of sensors. Choose those that enable you to control your lighting and security system most efficiently.

Other Materials

For wiring and fixture installation, you'll also need 12-inch wood stakes and a marking pen for marking fixture locations, electricians' tape, solvent glue for PVC conduit, cinder blocks and concrete for anchoring fixtures and receptacles (see page 80), a narrow roll of plastic sheeting for trenches dug through lawn or flowerbeds, caulking compound for sealing holes made in building walls, and the hardware for mounting boxes on the house or other structures.

Tools

Have the following tools on hand for the job: a hacksaw for cutting PVC pipe or rigid metal conduit; a rented conduit bender; several pairs of pliers, including needlenose, lineman's, and side-cutting types; Phillips and standard screwdrivers, a pipe reamer; a cable stripper; and a voltage tester. For a conduit system, you will need to rent or buy a fish tape for pulling the cable through the finished conduit. For larger installations of rigid metal conduit, you can rent a threader for threading the pipe ends. For trenching and cleanup, you probably already have a round-point shovel, a spade, a broom and rake; you can also get a narrow trenching shovel for faster work. For a large installation, you can even rent a trenching machine.

List all the materials and tools you need. Electrical parts, conduit, cable, boxes, fittings, and tools can probably be purchased at a local hardware or electrical retailer (some wholesale electrical supply houses may be willing to sell to you as well), but you may have to send for the lighting fixtures, time clocks, sophisticated switches, and sensors. See the list of sources on page 39 for anything you can't find locally.

SAFE OUTDOOR WIRING

Electricity can be life-threatening only if you don't respect it. Always turn off the power to a circuit you are working on, and be sure it really is off by checking with a voltage tester. And be extremely careful about working with electricity around water or moisture of any kind. Here are some additional suggestions:

1. Before starting to install your lighting, learn about your local electrical code from your city or county inspector, who is usually attached to the department of building inspection. You can also get a copy of the National Electrical Code, which forms the basis for most local codes, from the National Fire Protection Association, Batterymarch Park, Quincy, MA 02269. The NFPA also publishes a handbook which describes the code in greater detail. The electrical code is intended to protect you from the hazards of an improperly installed system, so know and follow its requirements. These include specifications for the types of conduit, wire, cable, and other materials allowed for outdoor wiring in your area; the depth of burial for different types of cable and conduit; the use of GFI's to protect outdoor receptacles and fixtures; and electrical wiring in or around water.

2. Get all required permits and inspections of your system as you design and install it. On 120-volt wiring, a permit is usually required for any new work or extension of your present service, but not for repairs and replacement. Twelve-volt wiring rarely requires a permit, but one may be needed for the wiring of the GFI receptacle into which you'll plug the transformer.

3. Never work on a live circuit. Before starting, remove the fuse or turn off the circuit breaker that controls the circuit and post a sign at the box telling others not to touch it while you're working. Turn off the circuit for any outdoor wiring work, including such simple procedures as changing the lamp in a fixture.

4. After turning off the power, make sure the circuit is dead by using a voltage tester. To test a receptacle, insert one probe into each slot; if the tester's lamp lights up, the circuit is live. To test a switch, remove the face plate, touch one probe to the metal box and the other first to one screw terminal and then to the other. If the box is plastic, touch one probe to the screw terminal and the other to the bare ground wire. If the lamp lights up when you touch either screw terminal, the power to the circuit is still on. Safety note on using a voltage tester: handle the wire leads only by the insulation when testing a circuit, and don't use it when standing on wet or moist ground.

5. The combination of moisture and electricity can cause you to get a severe shock. Don't work on a damp floor indoors or wet ground outside when doing wiring. Stand on a rubber mat or dry boards, or wait until the earth dries to do the work.

6. Use wire and cable sizes rated for the electrical load they will carry and the length of the wire runs in your system (see the table on page 70). On low-voltage systems, follow the manufacturer's recommendations.

7. Make sound electrical connections to avoid short circuits. Avoid damaging wires when pulling them through the conduit. Damaged insulation may not show up for several years, until dirt and moisture have accumulated. Strip only enough insulation from the wires to insert them into the wire nut or to wrap the binding screw two-thirds of a turn. Put all boxes for connections at switches, receptacles, and fixtures at least 12 inches above the ground.

8. Use only cable, conduit, boxes, and other materials approved for outdoor use by Underwriters Laboratories, or in Canada by the Canadian Standards Association.

9. Enclose any cable for 120-volt outdoor wiring in conduit, or bury direct-burial UF cable deeply to protect the wiring. Run the cable or conduit along natural landscape barriers, such as walks and header boards. You can safely leave low-voltage cable on the ground surface, but the garden will be more attractive and the wire better protected if it is buried at least six inches deep.

10. Leave the following jobs for an electrician: checking your final lighting plan for correct wire sizes, length of wire runs, and electrical loads; doing any wiring in or around swimming pools, garden pools, and other water elements; and wiring in remote-control relays on large 120-volt systems. If you are unsure of how to do the rough or finish wiring of any other component of your 120-volt system, you can find complete instructions in Ortho's book, *Basic Wiring Techniques,* for wiring switches, receptacles, new circuits at the service panel, and connecting the system to power. If you are not completely comfortable doing any of these jobs, leave it for the electrician to do.

STEP 3: *Mark the Locations of Fixtures, Receptacles, and Switches in the Garden*

Work from your lighting plan, using wooden stakes to mark each location. Note on each stake with a marking pen the fixture stock number or box type and any important information about its positioning. Then, beginning at the point where each circuit will connect to its power source, mark the wire runs on the ground with agricultural lime. Do this by dribbling the lime from your hand as you walk along the planned line of each trench. If you are connecting to an existing circuit, the power source may be an existing outdoor receptacle or switch. If you are installing new circuits or an additional service panel, mark trenches running to the location of the existing or new panel. Lay them out to be as straight as possible; if the trench must go around obstructions, plan a gradual curve rather than a sharp bend.

Mark wire runs on the ground with agricultural lime. Lines can be drawn over any surface, and erased and redrawn easily.

Above left: Before digging through a lawn, remove the turf and set it on a sheet of plastic.
Above right: Mark the fixture locations with a stake.
Left: This metal conduit bender can be rented.

STEP 4: *Dig the Trenches*

If you are trenching through a lawn, remove the sod with a spade to a depth of 2 inches. Cut it into uniform rectangular pieces and put them aside on plastic sheeting. If you have much sod to cut through, rent a sod cutter; it will take the sod up in neat rolls. Keep the sod moist during the project, and replace it as soon as possible. If you must trench through an established flower bed, lift out perennials and small shrubs carefully and put them aside in a shallow ditch, pulling soil up to the roots and keeping them moist. Once the trench line is cleared, excavate the trenches to the depth required by the code, using the trenching or round-point shovel. Make trenches at least 4 inches wide.

If there is no "sleeve" of pipe where the conduit or cable must go under a walk, dig the trench up to each side of the walk and drive an iron pipe under the walk with a sledgehammer. Pull out the pipe and push the conduit or cable through the hole it leaves. Or rent a drill with an earth boring bit 10 or 20 feet long.

STEP 5: *Lay Out the Materials Along the Trenches*

Place the conduit, fittings, boxes, and fixtures near where they are going to be installed. If you are using metal conduit, cut it into the required lengths with the hacksaw and smooth any burrs on the cut end with the pipe reamer. Bend the necessary curves with the conduit bender. Avoid making the curves too sharp or crimping the pipe, since this will make it hard to pull the cable through later. With PVC conduit, simply cut it into sections with the hacksaw, smooth the burrs, and lay the pieces along the trench. For curved sections of PVC conduit, purchase the specially-bent pieces of pipe called sweep bends. Lay the threaded PVC couplings for fitting to the boxes next to each box location.

STEP 6: *Assemble the Conduit*

If your outdoor wiring will be enclosed in conduit, assemble it now, before pulling the wire through. To put together rigid metal conduit, thread any pieces you need to with the threader (full-length pieces are sold threaded on both ends, and pieces attaching to compression couplings don't need threads), then start at one end, putting together each piece of conduit with its fitting and adding the next piece. At the location of the upright section of conduit leading to each above-ground box, widen the trench slightly and slip a cinder block over the upright conduit, and set the block level in the bottom of the trench. After you've installed the fixtures and checked the system, you'll fill this with concrete to anchor the fixtures. Add the junction boxes and boxes for outdoor receptacles while connecting the conduit, as you come to them. Use approved outdoor boxes or below-grade outdoor boxes in all cases except when installing switches indoors; here use standard indoor boxes. Don't install the fixtures yet.

Assemble PVC conduit by first daubing the inside of the fitting, then the outside end of the pipe, with a light coat of glue, then push them together. To spread the glue, twist the pipe about one-quarter turn as you insert it into the fitting. The glue sets up in just a few seconds, so work quickly. Drop the pipe into the trench, place the cinder-block anchors, and install the boxes as above.

If you are using above-ground boxes with a PVC system, use steel conduit at least above ground level. For durability, extend the steel conduit through the cinder block anchor and make the transition to PVC at the 90-degree bend.

Even if you are directly burying UF cable, you must use conduit on the above-ground portions of the system. This means enclosing the cable in conduit from any above-ground junction box, receptacle, or fixture, down at least one foot into the ground. Check your local code for the extent to which conduit is required near these connections.

It is usually best to put electrical conduit in its own trench. If for some reason you are installing it near a water or gas line of a similar material, mark the electrical conduit at intervals with spray paint to distinguish it.

Top: A concrete block (which will later be filled with cement) provides a solid anchor for each fixture.
Bottom: Sweep bends make smooth curves in plastic conduit for easier wire pulling.

STEP 7: *Pull the Wires Through the Conduit; Run Direct Burial Cable and Tree Cable*

Working from your lighting plan, run the wire for each of your lighting circuits in turn. Beginning with the first circuit, run the wire from the power source (without connecting to it) to the location of the new switches that will control that circuit. Leave several feet of wire at the power source to make the connection. If the switches are indoors, choose the route that causes the least damage to the inside walls and ceilings; this will often be through an unfinished basement or attic, and may involve drilling holes through floor or ceiling joists and fishing the cable through walls (see below for instructions for using a fish tape). Use conduit or armored cable indoors if your code requires it. Run one cable from the power source to the location of each switch, then short pieces of wire or cable, called "pigtails," to the next switch on the circuit. From the location of each switch, run a second cable out to the lighting fixtures and receptacles it controls. For most lighting systems, use 2-wire cable with a ground wire, or in some special cases (such as in wiring a 3-way switch), 3-wire cable with a ground. See the table on page 70 for assistance in determining the size of wire you need on each circuit.

If your outdoor wiring will be installed in conduit, pull the cable coming from the switch location through the already-assembled conduit. The best tool for pulling wire through both interior house walls and conduit is a fish tape. These can often be rented at local tool rental stores. Pull the cable through each section of the conduit in turn; cut the cable at each box and leave several inches extra to make the connection. On systems of direct-burial cable that only use conduit above ground, coil an extra foot or two at the base of the upright section of conduit, beneath each receptacle or fixture. Leave extra wire for fixtures hung in trees also. This makes relocation easier should it be needed later.

STEP 8: *Connect the Wires to Switches, Receptacles, and Fixtures*

Once you have run the cable to the indoor boxes for switches and the outdoor boxes for receptacles and lighting fixtures, you can complete the wiring yourself, following the instructions in Ortho's book, *Basic Wiring Techniques,* or have an electrician do it for you. Before you begin wiring, find out when the inspector will want to inspect your work. The inspector may want to see the wire splices before the final installation of switches and fixtures, or perhaps only the finished installation.

Connecting Switches

If your lighting will be controlled by a single switch or plugged into a receptacle that is operated by only one switch, it can be wired as a single-pole switch. If it will be controlled by two separate switches, wire each of them as a three-way switch. If you have found it most convenient to run the wire through the fixtures before it reaches the switch, it will be wired as a switch loop. For installations without conduit, begin wiring at each switch box by stripping the plastic coating from the cable where it enters the box and attaching it to the box with the built-in clamps or snap connector. Strip $\frac{1}{2}$ inch of insulation

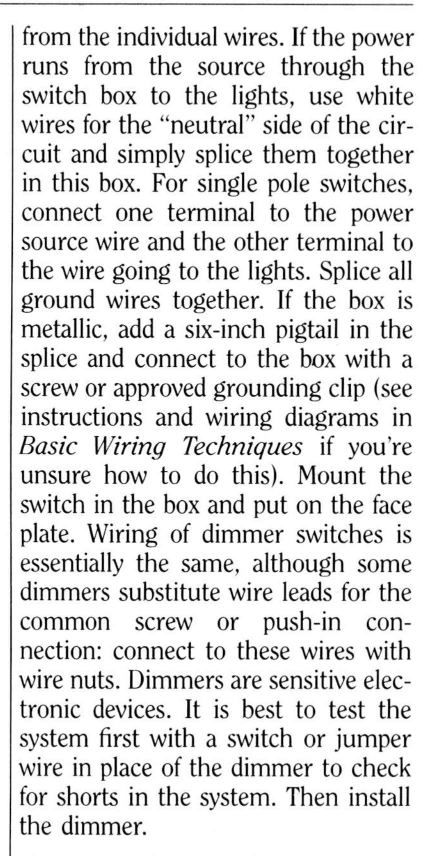

from the individual wires. If the power runs from the source through the switch box to the lights, use white wires for the "neutral" side of the circuit and simply splice them together in this box. For single pole switches, connect one terminal to the power source wire and the other terminal to the wire going to the lights. Splice all ground wires together. If the box is metallic, add a six-inch pigtail in the splice and connect to the box with a screw or approved grounding clip (see instructions and wiring diagrams in *Basic Wiring Techniques* if you're unsure how to do this). Mount the switch in the box and put on the face plate. Wiring of dimmer switches is essentially the same, although some dimmers substitute wire leads for the common screw or push-in connection: connect to these wires with wire nuts. Dimmers are sensitive electronic devices. It is best to test the system first with a switch or jumper wire in place of the dimmer to check for shorts in the system. Then install the dimmer.

Connecting Receptacles

Install outdoor receptacles only in boxes rated for outdoor use, to protect them from moisture. Be certain that these and all outdoor boxes are at least 12 inches above ground, unless they are special cast-iron boxes designed

Left: Fish tape is used to pull the wire through the conduit. Right: Wire nuts thread over the ends of the wires to make connections.

Left: Thread the wires through the junction box.

Right: Connect the wires for each fixture according to the manufacturer's instructions.

Above: Connect the wires in the junction box.

for below-grade installation. Connect the receptacles as you would indoor receptacles. An end-of-the-run receptacle will be connected only to two incoming wires, a middle-of-the-run receptacle to two incoming wires and two that go on to the next receptacle or fixture. Remember that the first receptacle on an outdoor circuit should be a GFI receptacle, unless the circuit is protected by a GFI circuit breaker at the service panel.

Attaching Fixtures

At each middle-of-the-run connection for a fixture, the junction box should have two threaded holes in the bottom for the incoming and outgoing conduit and one threaded hole in the top, in which the fixture is usually mounted. The box for each end-of-the-run fixture will have only two holes: one for conduit and wire coming in, the other for the fixture itself. Wire the fixtures according to the instructions that came with them, and screw on the box cover. If you are installing a 120-volt fixture attached to a stake and flexible cable, use the junction box recommended by the supplier and be sure that a waterproof gasket protects the box from moisture at the opening for the cable. If you are installing high-intensity discharge fixtures and lamps, follow the manufacturer's wiring instructions. For outdoor work, always point the open end of the wire nut downwards after tightening to prevent water from collecting and corroding the connection.

STEP 9: *Have an Electrician Do Any Wiring Around Water*

Hire a licensed electrician to do any installation of cable, conduit, fixtures, or receptacles that will be underwater in a pool, or within 10 feet of any garden water element. The electrical code is very demanding in this area, and the risks of poor wiring are too great to do this work yourself, unless you are a skilled electrician.

STEP 10: *Have Your Wiring Checked and Connect To Power*

If your system will not be inspected by the building department, have it checked now by the electrician for proper burial depth of conduit or cable, good electrical connections and grounding, and correct materials.

When the system has been approved by the building department, you can connect it to its power source yourself, following the directions in *Basic Wiring Techniques,* or have the electrician do this for you. Be sure to follow proper safety precautions. Turn off the power before making the connection, and check that it is off with a voltage tester. Follow the other precautions on page 78.

STEP 11: *Test the System*

Turn on the switches one by one and make sure all the fixtures work. If you find one that doesn't, try replacing the lamp first. If this doesn't solve the problem, turn off the power and check with a voltage tester to be sure it's off. Then check the connections and the switch itself with a continuity tester (see *Basic Wiring Techniques* on how to test switches and receptacles).

Check also the position and aiming angles of fixtures. This can best be done at night, and you will probably find small adjustments you want to make. You may even need to shift a fixture a short distance, necessitating turning off the power, disconnecting the fixture, digging a new trench, rerouting the conduit, and reconnecting the fixture. This takes some self-discipline when you are so close to the end of the job, but it is better to do this work now than to live with a lighting effect you don't want.

Once you are satisfied that the system works and does what you want it to, anchor each fixture and free-standing receptacle firmly by pouring ready-mix concrete into the cinder block that the conduit passes through.

STEP 12: *Backfill the Trenches and Clean Up*

Once the inspector has made the final inspection, begin the cleanup. Backfill any trenches through lawn areas carefully by filling them halfway to the top with earth, flooding the trench with water, then filling the rest of the way and flooding again. Then replace the sod. This process guards against later settling of the sod as loose backfill settles in the trench. Fill around underground conduit and cable with care, keeping rocks or other sharp objects from touching the cable. Tamp the trench down well.

Caulk around any cable or conduit that passes through the exterior house walls, and around outdoor receptacles and switches attached to the house.

INSTALLING TWELVE-VOLT LIGHTING

It is much simpler to wire a low-voltage lighting system than a standard (120-volt) system. Twelve-volt systems usually don't need permits and inspections, are less dangerous to work with, don't require deep trenching, and have such low power needs that you may not need any new circuits to operate them. It is true that some inexpensive low-voltage systems may not be very durable and that the levels of illumination they give are less bright than standard-voltage lighting, but you can find low-voltage systems of excellent quality that will meet many lighting needs, especially in small gardens.

Flooding the trench will settle the soil before the sod is replaced.

Despite its relative safety, 12-volt lighting can be hazardous if poorly installed. Loose connections at the fixtures or too many fixtures on a cable can cause heat to be generated, and possibly fire. Follow the manufacturer's instructions carefully when running the cable and connecting fixtures.

The installation process for low-voltage lighting parallels that for standard-voltage systems. First, find out whether any permits are required, then order the materials, lay them out, run the cable, connect fixtures and transformers, test it, and clean up.

In installing 12-volt lighting, you may need to work with 120-volt wiring to the extent that you need new outdoor GFI receptacles or indoor switches to control the system. Follow the safety rules on page 78 if you do the work yourself, or hire an electrician to do it for you.

If you are putting in low-voltage lighting as part of a new landscape, first design the system, do any 120-volt wiring for the transformers, and lay sleeves for the cable beneath any proposed walks or patios before the landscaping begins. When all the outdoor construction is finished and all the plants in except the lawn, ground cover, and bedding flowers, put in the lighting cable and fixtures. Then finish planting and clean up.

STEP 1: *Get Permits, If Required*

Permits are rarely required by local building codes for a 12-volt lighting system. If, however, you are installing new 120-volt circuits to operate a large low-voltage system, or a new GFI receptacle to plug in a transformer, check with the building inspector to see if you need a permit. Find out at the same time if your local code lists any requirements for low-voltage systems.

STEP 2: *Order Materials*

Materials for low-voltage lighting may be hard for you to find locally. Your hardware store may have some low-voltage systems available in kit form, as well as many of the basic wiring materials required, such as direct-burial cable and wire nuts, but other parts and fixtures will have to be ordered. Several of the manufacturers listed on page 39 offer low-voltage systems and materials; you may already have ordered their catalogs when designing the system. Be sure any materials you buy locally meet the specifications of the low-voltage manufacturer.

Cable
Low-voltage cable is usually two-conductor, direct-burial cable of wire sizes number 12, 14, or even 16. The easiest way to get the correct wire for your system is to order it with the fixtures from the manufacturer. Follow the manufacturer's recommendations for maximum length of wire runs in your system.

Wire Connectors
Some systems come with their own clamp-on terminals, which can be fit without cutting the cable. Other fixtures are connected with wire nuts, which are usually supplied with them.

Switches and Receptacles
Order any new outdoor 120-volt receptacles you need to power your transformers. These should either be GFI receptacles or installed on a circuit protected by a GFI. Whether your transformer will be plugged into a receptacle on a 120-volt circuit or wired directly into the circuit, it should be controlled from indoors by a switch. This can be a standard wall switch or a time clock or wired-in time switch, or the transformer can be plugged into a special receptacle module and operated by a remote control plug-in switch (see page 41).

Fixtures
Most of your low-voltage fixtures will be set into the ground on spikes or hung in trees, but for special situations you can order some with a deck or patio mount. Other accessories available are shields to prevent glare, mounts for installing fixtures on conduit, and various lenses to protect the lamp or color the light. Be conservative in the use of colored lenses; they create strong visual effects in the garden (see page 43).

Low-voltage fixtures are usually sold complete with the lamps. You may be able to specify any one of several different lamps for the same fixture to get brighter light, a broader floodlight, a narrower spot, or other effects, so check the catalog information carefully.

Transformer
Order the size transformer you need for your system. Low-voltage transformers are available to operate systems using up to several hundred watts. You can divide up larger systems among several transformers. Order your transformer for indoor or outdoor installation; those for indoor installation are less expensive, but some require the wiring in of a fuse for installation. Some manufacturers combine a transformer with a time clock in the same housing; others sell them as separate units. Most come with just two screw connections for the low-voltage cable to the fixtures, but you can also order multiple terminals so that you can make several wire runs from the same transformer. Just be sure not to exceed the maximum wattage recommended for the transformer you select.

Time Clock
If the time clock is a separate unit from the transformer, you can plug the transformer into the receptacle built into the time clock, and the time clock into a standard receptacle, to operate the system automatically. You can also wire the transformer to any one of the many indoor time switches available (see the list of manufacturers on page 39). Most time clocks can only operate one lighting system, so all your lights will go on and off at the same time unless you wire different

The transformer and timer for a 12-volt system should be in an accessible location.

parts of your system to different time clocks.

Other Materials
Before you begin to install your system, you also need to gather 12-inch wooden stakes and a marking pen for locating the fixtures, agricultural lime for laying out the wire runs, and plastic sheeting for trenches dug through any delicate areas of the garden.

Tools
Also get three pairs of pliers: needlenose, lineman's, and side-cutting types; standard and Phillips screwdrivers; a spade for trenching; and a rake and broom for cleanup.

STEP 3: *Lay Out the System in the Garden*

Once you have on hand all the parts of the system, and the other tools and materials needed, take out your final lighting plan and lay out the system. As with 120-volt systems, start by driving a wooden stake at the location of each fixture and writing on the stake the name and stock number of the fixture, with any notes about its aiming angle and position.

Choose a location for the transformer. It can be indoors in a garage or closet, or outdoors, but it should be central to the lighting fixtures so that the wire runs extend out from it like the spokes of a wheel. This will keep the wire runs as short as possible.

Mark the wire runs with agricultural lime. When possible, try to follow "natural barriers" in the landscape such as header boards and path edges, to protect the wire from accidental damage. Do not make the wire runs longer than the maximum length recommended by the manufacturer, to avoid voltage drop and poor brightness of the lamps at the end of the runs. If you must exceed the maximum length, you will have to use a larger wire size or two cables wired in parallel. Contact the manufacturer or your supplier for assistance with this.

STEP 4: *Run the Cable*

You can safely lay low-voltage lighting cable right on the ground surface between the fixtures, but your garden will be more attractive if the cable is hidden several inches below ground. Begin at the transformer end of each wire run and make a narrow "slit" trench 6 inches deep with a spade, along the chalk line marking the run. Drop the cable into the bottom of the trench, leaving a foot of it coiled at the transformer and a loop above ground at the location of each fixture, and continue along the wire run to the last fixture.

Run cable into trees for any fixtures to be installed there. Use a metal staple and a plastic strap for attaching cable to each tree. Try to place the cable so it won't be highly visible. Also install now any other above-ground sections of cable, such as those to deck or arbor fixtures. Here, too, try to hide the cable from view. Attach it to wood with electrical staples, to concrete with small conduit straps mounted with lead or plastic screw anchors in the concrete.

STEP 5: *Connect the Fixtures*

It's important to connect the fixtures to the cable before plugging in the transformer. Otherwise, you can expect to blow the transformer fuse or trip its small circuit breaker. Check the locations of the fixtures again with the plan, then connect the cable to each according to the manufacturer's instructions. You can connect to some fixtures with simple clamp-on connections which save you cutting the cable. For others you must cut the

Top: Use plastic insulators to attach cable to trees.
Bottom: Connections for 12-volt systems do not require a junction box.

cable and make the connection by conventional methods, using wire nuts. Most low-voltage fixtures are packaged with a wiring diagram enclosed. The connections usually should be made above ground and should be protected in some way from moisture by enclosing the connection in the hollow stem of the fixture or sealing it with an electrical sealant. If connections are made underground, seal them in an epoxy packet (this and the sealant mentioned above are available from electrical distributors).

When all fixtures and cables are connected, check to be sure all the fixtures are vertical and aimed correctly.

STEP 6: *Install the Transformer*

Your transformer should either be wired directly to an indoor switch or connected to a GFI-protected receptacle that is, in turn, operated by an indoor switch. If the receptacle is outdoors, it should be of a weatherproof design. Alternatively, plug the transformer into the receptacle of a separate time clock and the clock into the 120-volt receptacle. If you are installing several lighting systems and transformers, locate the switches indoors in one central location.

Attach the low-voltage cables from the wire runs to the transformer, and plug it in.

STEP 7: *Test Your Lighting System*

This is best done in the evening or at night, when you can see the lighting effects. Once the fixtures are on, check each for proper placement and aiming angle. One wonderful thing about low-voltage systems is that you can easily relocate fixtures if you need to.

If any fixture does not go on, check the lamp first; if it is good, turn off the transformer and check the connection to the cable. Make the connection again if necessary. If all the fixtures work, but those at the end of the line are dim, check the length of the cable run to see that it doesn't exceed the manufacturer's recommended length. If it does, the dimness is being caused by voltage drop, and can be remedied by rearranging the wire runs so none exceed the maximum length.

Connections will be pushed into the post before it is buried.

COMBINING 120-VOLT AND 12-VOLT LIGHTING

Both standard-voltage and low-voltage lighting have distinct advantages, as discussed in the chart on page 29. You may decide to install both types of system in your garden, for more lighting flexibility and beauty. Perhaps you'll use low-voltage lighting in small, intimate garden areas, or for soft, general lighting of the garden, while wiring the security lighting or porch fixtures to 120-volt power.

If you do choose to install both types of lighting systems, follow these guidelines:

1. Choose fixtures for the two systems that match as well as possible. Some manufacturers make both types of system, and these may be good sources.
2. Wire the systems separately, according to the instructions in this chapter. Each type of system will serve particular purposes: the low-voltage lighting will add a gentle night beauty to the garden; standard-voltage lighting will give brighter illumination where it's needed. With separate wiring and switching, you will have the greatest lighting flexibility.
3. Use small lamps in the 120-volt fixtures. Ideally you should use no 120-volt lamp larger than 75 watts. This will avoid overpowering the low-voltage lighting when both systems are on.

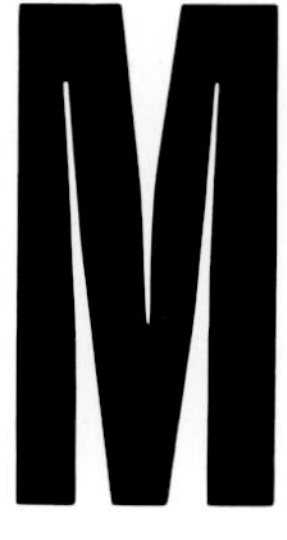

AINTAINING YOUR LIGHTING SYSTEM

Because you have designed your lighting carefully and have installed the best system you can, you'll naturally want to protect your investment with proper maintenance. Lighting system maintenance isn't difficult and need not take much time.

Your maintenance program begins during the design process, with the choice of appropriate materials for your system. After the lighting is in, you will need to keep the fixtures clean, free from corrosion, and adjusted properly, and you will need to periodically replace lamps. Occasionally, you may have to deal with more difficult problems; the troubleshooting suggestions on page 89 will help you with these.

Planning for Maintenance

You should begin thinking about the maintenance of your lighting system during the design process, when you are choosing the lighting fixtures, switches, receptacles, boxes, transformer, and time clock. Study the product catalogs and ask questions of your electrical supplier or hardware staff to find out which lighting products will last the longest with fewest problems. You will find that the best lighting materials will cost more, but if you can afford them, they will save you maintenance hours and dollars. For example, you will pay more for fixtures of heavy-gauge aluminum, copper, or bronze, for fixtures with porcelain lamp sockets, or for cast-iron underground junction boxes, but they should last longer than fixtures made of lightweight plastic with thin metal sockets and cast-aluminum boxes.

When you are designing the system, there are several other ways you can minimize maintenance. A major maintenance task is the changing of lamps. The high-intensity discharge lamps, which include mercury vapor, metal halide, and high-pressure sodium, have a much longer life than incandescent lamps. They are also more expensive, but can be useful in systems that will be on many hours each night or in particular fixtures that will be inaccessible for frequent lamp changes, such as those at the tops of large trees. If you are using types of incandescent lamps designed for use indoors (Types A, R, and ER lamps), you can extend their life by installing them in fixtures that protect the lamp with a lens or enclose it so that rain and sprinkler water will not hit it. You can also design circuits of incandescent lamps with dimmer switches in order to precisely control the lighting intensity. These can extend the life of the lamps if you use them to provide only as much light as you need.

Although fixtures high in trees cast interesting shadows, their lamps are difficult to change.

Intended Use

Order and use all fixtures, cable, boxes, and other materials only for their intended use. This means being sure that all switches and transformers are designed for outdoor use, putting only special below-grade type junction boxes in contact with the soil, and using the proper size of cable on each wire run. Check the manufacturer's specifications when buying the product, and your local code for the recommended uses of various cables, boxes, and switches. This will save you a great deal of maintenance trouble later and assure you that each fixture will perform as it's designed to for a long time.

A well designed and carefully installed lighting system should provide years of enjoyment.

Cleaning

Keep dirt, leaves, and other debris out of outdoor lighting fixtures. This is especially important with the fixtures used for uplighting, because debris can rapidly accumulate around the lens or lamp. Periodically, you should also wipe the lamps and fixtures with a damp cloth to remove dust and deposits from sprinkler watering. Do this only when the lamps are cold. Clean underwater fixtures to remove mineral deposits by first scraping off the thickest deposits with a razor blade, then scrubbing them with a solution of 1 part muriatic acid to 8 parts water. Wear rubber gloves for this job.

Corrosion and Lubrication

If you live near the ocean, are locating fixtures near roads that are salted in winter, or for any reason expect problems with corrosion, choose the most durable fixtures you can get. Otherwise, some parts of your lighting system may have a disappointingly short life. These durable materials include copper, bronze, and cast-aluminum fixtures with porcelain sockets, cast-iron junction boxes, and rigid PVC conduit (rigid metal conduit tends to corrode more rapidly).

Even if you don't expect unusual amounts of corrosion, you should remove all lamps from their fixtures once each year and spray the sockets with a silicone-based lubricant.

Pruning and Relocation

For full light output, keep shrubs and ground covers pruned back from lighting fixtures. The exceptions to this are tree-mounted fixtures; you can allow some leaves and even small branches to grow in front of these, because it will help diffuse their bright light and eliminate glare.

When trees in which fixtures are mounted do need pruning, be careful not to jar or otherwise damage the fixtures. After a major pruning, you will probably need to relocate some fixtures so that the light is again shielded by the leaves or branches and no brilliant glare is seen.

For cables that were installed in trees or large shrubs, you will periodically need to reattach the metal staple and plastic strap.

Relocation of ground-mounted 120-volt fixtures is usually unnecessary if your design has been carefully thought out. If you do need to move one of these fixtures, turn off the power, then dig carefully down to the base of the conduit. If the fixture is wired with UF cable and the conduit extends only a few inches into the ground below the fixture, dig down to the several feet of cable that you coiled at the base of the conduit during installation. Then dig a trench to the new location. If this is within a few feet, you can move the cable, section of conduit, box, fixture and cinder-block anchor to the new location, and reposition them. This will be especially easy if the fixture is at the end of the run. If it is in the middle of the run, if it must be moved farther than the existing cable allows, or if all the wiring is enclosed in conduit, relocation will be more difficult. To move a middle-of-the-run fixture, turn off the power, dig out around the fixture and the fixtures on both sides of it in the series. Run conduit and wire to the new fixture location from the fixtures on both sides of it in the run. Disconnect the wiring to the old location, move the fixture and box to the new location, put in a new cinder-block anchor, and wire the fixture.

Some 120-volt fixtures have built-in adjustments for changing the position of the lamp. These include adjustable well lights and uplighting and downlighting fixtures with adjustable swivels. These fixtures can be checked periodically and re-aimed as you wish to change the lighting focus. Even more versatile are those 120-volt fixtures with a portable stake mount, which can be moved several feet in any direction from the junction box.

It is easier still to move 12-volt fixtures. Low-voltage systems are designed to be portable; relocating a fixture is especially easy if the cable has been laid on the ground surface: simply pick it up and move it. If the cable doesn't reach to the new spot, splice additional cable to it, sealing the connection in an epoxy packet. You can then bury this splice, but it is better to leave it on the surface so it is easily found if there are problems. Do at least hide it behind a plant, however. To relocate a fixture where the low-voltage cable has been buried, dig up the cable, uncoil the several extra feet you left beneath the fixture, and move it.

Low-voltage fixtures frequently get out of adjustment or position because most of them are anchored only by ground stakes, with no conduit to provide additional support. Check the fixtures periodically, reposition them if needed, and stand upright any that are leaning.

Appendix

TROUBLESHOOTING

120-Volt Systems

Many problems with standard-voltage systems are best investigated and remedied by a professional electrician; a few may be easy for you to solve. Remember to work safely. Always turn off the power before doing any work on a circuit, and make sure it is off by checking with a voltage tester. Do not work on a wet or damp surface; stand on a rubber mat or dry boards to avoid this if necessary, and make sure your hands are dry before touching the wiring. This is especially important when testing a live circuit with a voltage tester. When using the voltage tester to check a connection, handle it only by the insulation, never by the bare probes. Don't touch any metal in the box or on the wall while you are testing the circuit. Refer also to the safety rules listed on page 78 before starting any repair.

Problem:	Solution:
Entire system won't light	1. *Fuse blown or circuit breaker tripped.* Turn off the power and replace the fuse or reset the circuit breaker. If it blows or trips again, see below (under "fuse blows, circuit breaker trips").
	2. *Bad switch or connection to power.* Turn off the power, making sure it's off with a voltage tester (see page 78 for method of use). Open up the junction box or switch that is your power source, pull out splices, and test connections with the voltage tester (one probe on neutral white wires, one on hot black wire connection). Follow the precautions listed above. If there is no current at connection, turn off the power, untwist wires, and reconnect them. Test again. If there still is no current, the problem is further back in the circuit. Consult an electrician. If the connection is good, the fuse or breaker did not blow, but there still is no output to fixtures, have the system checked by an electrician.
One fixture won't light	1. *Lamp is bad.* Turn off the power. Remove the lamp and test it in a fixture you know works. If the lamp is bad, replace it. If it still doesn't light, replace it with another new lamp. (You may have gotten two bad ones in a row.)
	2. *Lamp-socket contact faulty.* If the lamp works in other sockets but the fixture still won't light, turn off the power again. Remove the lamp, turn on power, and test socket with the voltage tester. Follow the safety precautions given above. If the socket is good, turn off the power, bend the spring terminal slightly upward with a screwdriver or finger to improve the contact with the lamp, replace the lamp, and test.
	3. *Junction box wiring at fixture is faulty.* If there is no power to the lamp socket, turn off the power. Test at the receptacle or switch with the voltage tester to be sure it's off. Open the box, pull out the splices, and check the connections with the voltage tester. If connections are good, remove the fixture and return it for testing or replacement.
Fuse blows, circuit breaker trips	1. *Loose fuse.* Make sure the fuse is screwed in tightly.
	2. *Short circuit.* Check the condition of a blown fuse. If the inside of the glass is darkened, or if the fuse blows or breaker trips immediately when the circuit is turned on, suspect a short circuit. To locate it, turn off the power, unplug all cord-connected fixtures and equipment from receptacles on the circuit, and turn on the circuit. If it blows again, the problem is in the fixed wiring. Follow troubleshooting techniques in next paragraph or consult an electrician. If it does not blow, turn off the power again, reconnect the first fixture, and turn on the power to see if it blows. If not, add other fixtures and equipment back to the circuit one by one, turning the power off each time, connecting the fixture, and turning it back on to see if the breaker trips or fuse blows. When you find the problem fixture, turn off the power, check connections, and rewire or replace the fixture as necessary. The short circuit may be in permanently wired fixtures, receptacles, or in underground wiring. Turn off the power. Remove covers from junction boxes and receptacle boxes, looking for blackened connections or accumulations of dirt and water. If connections appear clean and dry, disconnect lighting circuit by opening the connections in first junction box. Turn on the power. If it blows again, the problem is not in the lighting and you may wish to have an electrician do the troubleshooting. If the circuit stays on, repeat this process down the chain of junction boxes: turn off the power, reconnect the splices, go to the next box and open splices, and turn on the power. In this way, work through the system until the defective link is found. Shorted underground wire must be replaced. Shorted wire in conduit can be repulled using the existing wire, if dirt has not accumulated heavily in the pipe. Water accumulation in the pipe is common and is not a problem as long as the wire is installed carefully. If the wire will not pull, it is either jammed in the pipe or is a direct-buried cable. The entire assembly must be dug up and replaced.

Problem:	Solution:
	3. *Overloaded circuit.* If the fuse does not blow or breaker trip immediately when circuit is turned on, but does after several minutes, suspect an overload. An overload is also indicated by the appearance of the blown fuse: the glass is clear and the separated metal strip is visible inside. List all fixtures and other appliances on the circuit and add up wattages. Compare the sum to the maximum allowable wattage on the circuit (1500 watts on 15-amp fuse or circuit breaker, 1800 on 20-amp). If the sum is over the circuit's capacity, put some fixtures on another circuit with greater excess capacity, so that the original circuit is within allowable maximum. If the total wattage is within capacity, check for a short circuit by the method given under 2, above.
GFI trips	1. *GFI is faulty or incorrectly wired.* Check the product's directions for test procedure, turn off the power, disconnect the lighting circuit, and test. If it tests OK, turn off the power, reread wiring instructions for the GFI, and check for proper wiring.
	2. *Ground fault in system.* If the GFI is properly wired and tests good, check for a ground fault in the system. Turn off the power and disconnect all fixtures and receptacles but one. Turn on the power. If the GFI trips again, the problem may be in the wiring of this fixture. Check it and rewire as necessary. If the GFI does not trip, add fixtures back to the circuit one by one, testing as for short circuit until you locate the problem. If you cannot find a faulty connection but the GFI continues to trip, call an electrician.

12-Volt Systems

Problems here may occur in the house wiring that serves the transformer, in the transformer itself, or in the low-voltage wiring. An electrician can help with problems in the 120-volt circuitry. On the rest of the system, you can probably find the problem yourself, and fix the wiring or replace the faulty equipment as necessary. Remember to turn off or unplug the transformer before disconnecting or reconnecting a fixture.

Problem:	Solution:
Entire system won't light	1. *Blown fuse or tripped circuit breaker on transformer.* Check the fuse or circuit breaker on the transformer. If it's blown, see below ("fuse blows or circuit breaker trips") for solution.
	2. *Blown fuse or tripped circuit breaker on 120 volt circuit.* The problem is in the 12-volt circuit. Check the 120-volt connections or call an electrician.
	3. *Poor connection to 120-volt power source.* If both 12-volt and 120-volt fuses or breakers are good, turn off the 120-volt power, check to be sure it is off with a voltage tester, and test the connection to power source. If the power source is a receptacle, check it with any plug-in appliance. If the connection at the junction box or receptacle is faulty, rewire it or call an electrician.
	4. *Faulty transformer.* If power connections and fuses or circuit breakers are all good, but there is still no lighting, check the 12-volt output leads on the transformer with a voltage meter that will measure 12-volt current. If there is no 12-volt output, return the transformer for service or replacement.
One fixture won't light	See this problem above under 120-volt systems and handle similarly.
Fuse blows or circuit breaker trips when system is turned on	1. *Problem in transformer.* Disconnect the transformer from 120-volt power. Disconnect the circuit low-voltage cables from the transformer. Replace the fuse or reset the circuit breaker. Turn on the power to the transformer. If the fuse or breaker doesn't blow, the transformer is OK. If it does blow again, return the transformer for checking.
	2. *Short circuit in low-voltage cable connections.* If the circuit breaker doesn't trip or the fuse blows when all the low-voltage output cables are disconnected from transformer and power turned on, the likely problem is a short circuit in one of the output cables. Test them one at a time. First, with the power off, disconnect all the cables but one. Turn the power back on, and see if this cable causes the circuit breaker to trip. If it does, the short circuit is in this wire; if not, turn off the power, disconnect this cable, connect another, and turn on the power to repeat the test. Do this until you find the cable with the short circuit. Then find the exact location of the short circuit by the method described under "short circuit" for 120-volt systems, above.
Fuse blows or circuit breaker trips after several minutes of operation	1. *Overloaded transformer.* This is the probable explanation. Check for an overload by adding up the wattages of all the low-voltage lamps on the circuit, and compare the total to the capacity rating of the transformer. If the total wattage exceeds this rating, redistribute some of the fixtures to circuits supplied by other transformers with excess capacity. Or bring the total wattage down by replacing all the lamps with smaller lamps that use less wattage.

Problem:	Solution:
	2. *Short circuit.* If the total wattage on the transformer does not exceed its capacity, check the wire runs connected to this transformer for a short circuit, by the process given above.
Lights dim, especially at line end	1. *Voltage drop at end of run caused by overly long wire runs.* Check the length of the wire runs and compare them to the manufacturer's recommendations. If the runs are too long, redistribute fixtures or otherwise change the layout to shorten the wire runs.
	2. *Short circuit.* If the wire runs are not too long, check for a short circuit by the process given above.

PLANTS FOR LIGHTING

When lighting plants, study their form, branching structure, leaf and bark texture, reflectance, and seasonal changes. Each species can be lit to bring out its special beauty. Below, we list some common garden trees, shrubs, and perennials according to their dominant characteristics, and suggest ways to light them.

1. Broad, Dense Conifers
These trees can form the background for your garden, with their dark, solid presence. Many conifers are so dense that little light will penetrate their branches, so it's best to silhouette them by uplighting from the back or lighting a garden area behind them, or to uplight them with grazing light just touching the branch tips. You can plant and light a bed of white flowers or white-flowering shrubs in front of these trees for an effective contrast.

Some pines that are dense in youth mature to become tall, irregular, open trees. These can be lit with downlight from high in the branches; if they are growing in a grove, add occasional uplights for a striking effect.

White fir (*Abies concolor*)
Deodar cedar (*Cedrus deodara*)
Italian cypress (*Cupressus sempervirens*)
Colorado blue spruce (*Picea pungens* 'Glauca')
Many pines, including Austrian pine (*Pinus nigra*), Monterey pine (*P. radiata*), and Eastern white pine (*P. strobus*)
Western red cedar (*Thuja plicata*)

2. Large, Broad-leaved Shade Trees
If you have one or more of these trees in your yard, you're lucky. It can be the foundation of your lighting scheme. A large shade tree is an ideal place to mount fixtures high above the ground, hidden from below but lighting garden areas, paths, driveway, or patio. You can moonlight these trees or more simply downlight them. Add some uplight also to illuminate the leaf canopy above. Light a grove of trees with cross lighting from high in the branches.

Maples, including Norway maple (*Acer platanoides*) and sugar maple (*A. saccharum*)
Hackberry (*Celtis* species)
European beech (*Fagus sylvatica*)
Ash (*Fraxinus* species)
Many oaks (*Quercus* species)
American elm (*Ulmus americana*)

3. Trees with Attractive or Unusual Branching Pattern
Here are some of the best trees for moonlighting. Many have twisted, gnarled trunks and branches. Others, like the honey locust or Hankow willow, have an unusual branching pattern. Many of these trees are also attractive when uplit with a combination of floodlights and spotlights.

Honey locust (*Gleditsia triacanthos* var. *inermis*)
Olive (*Olea europaea*)
Chinese pistache (*Pistacia chinensis*)
Oaks, both deciduous and live oaks, including coast live oak (*Quercus agrifolia*), white oak (*Q. alba*), holly oak (*Q. ilex*), valley oak (*Q. lobata*), Northern red oak (*Q. rubra*), cork oak (*Q. suber*), Southern live oak (*Q. virginiana*)
Hankow willow (*Salix matsudana* 'Tortuosa')
Chinese elm (*Ulmus parvifolia*)

4. Trees with Colorful Foliage, Flowers, or Fruit
These trees can take over the spotlight in your night garden when the spring flowers fade. Several have attractively colored foliage all year, and can be focal points for permanent lighting. Others set heavy crops of fruit or berries that persist into winter, or have brilliant fall color. These can be lit seasonally, with a combination of uplighting and downlighting. Incandescent light brings out the reds and yellows of autumn leaves best.

Maple (*Acer* species)
Strawberry tree (*Arbutus unedo*)
Blue Atlas cedar (*Cedrus atlantica* 'Glauca')
Dogwood (*Cornus* species)
Hawthorn (*Crataegus* species)
Persimmon (*Diospyros kaki*)
Kaffirboom coral tree (*Erythrina caffra*)
Purple beech (*Fagus sylvatica* 'Atropunicea')
Ash (*Fraxinus* species)
Crape myrtle (*Lagerstroemia indica*)
American sweet gum (*Liquidambar styraciflua*)
Southern magnolia (*Magnolia grandiflora*)
Note: Downlight only, because uplighting lights brown leaf undersides unattractively.
Crabapple and apple (*Malus* species)
Black tupelo (*Nyssa sylvatica*)
Colorado blue spruce (*Picea pungens* 'Glauca')
Cherry plum (*Prunus cerasifera* 'Atropurpurea' or 'Thundercloud')
Chinese tallow tree (*Sapium sebiferum*)

5. Trees with Attractive Bark
The bark of some of these trees has a pronounced texture, which you can bring out with grazing light. Others have mottled bark, and can be effective focal points when uplit. Still others have highly reflective white bark, making them ideal for mirror lighting.

California buckeye (*Aesculus californica*)
Strawberry tree (*Arbutus unedo*)
Birch (*Betula* species)
Kaffirboom coral tree (*Erythrina caffra*)
American sweet gum (*Liquidambar styraciflua*)
Many pines (*Pinus* species)
London plane tree (*Platanus* x *acerifolia*)
Pomegranate (*Punica granatum*)
Many oaks (*Quercus* species)
Chinese elm (*Ulmus parvifolia*)

6. Shrubs with Glossy Leaves
The leaves of these shrubs have high reflectance. Downlight them to bounce light onto the patio or deck, or to make the garden shine with a play of brilliants.

Common camellia (*Camellia japonica*)
Mirror plant (*Coprosma repens*)
Weeping fig (*Ficus benjamina*)
Holly (*Ilex* species)
Glossy privet, Japanese privet (*Ligustrum lucidum*, *L. japonicum*)
English laurel (*Prunus laurocerasus*)
Shiny xylosma (*Xylosma congestum*)

7. Tall, Dense Shrubs
These shrubs can form the backdrop for a night view of the garden. For perspective lighting, plant them at the end of a garden axis and silhouette them by lighting behind them. Or emphasize their texture by grazing them with uplight or downlight. Light unevenly with a combination of spots and floods aimed from two sides to bring out their depth.

Privet (*Ligustrum* species)
Bayberry (*Myrica pennsylvanica*)
Sweet olive (*Osmanthus* species)
Pittosporum species
Carolina cherry laurel, Catalina cherry (*Prunus caroliniana*, *P. lyonii*)

8. Shrubs with Colorful Leaves, Branches, or Fruit
Place low area fixtures or accent uplights near these shrubs to highlight them in late summer and fall. Downlight massed plantings from nearby trees.

Red-leaf Japanese barberry (*Berberis thunbergii* 'Atropurpurea')
Red-osier dogwood and golden-twig dogwood (*Cornus sericea* and *C. s.* 'Flaviramea')
Smoke tree (*Cotinus coggygria*)
Burning bush (*Euonymus alata*)
Border forsythia (*Forsythia* x *intermedia*)
Bigleaf hydrangea (*Hydrangea macrophylla*)
Star magnolia, kobus magnolia, saucer magnolia (*Magnolia stellata*, *M. kobus*, *M. soulangiana*)
Pomegranate (*Punica granatum*)
Firethorn (*Pyracantha* species)
Rhododendron and azalea
Bridal wreath (*Spiraea prunifolia*)
Lilac (*Syringa* species)
Viburnum species

9. Perennials for Accent Lighting
These dramatic plants can be the focal points of a small garden. Several of the plants listed have large, glossy leaves. The foliage of others is sword-shaped. Both types can be lit by shadowing or silhouetting. The flowers of delphinium, globe thistle, and foxglove can be silhouetted or lit by low border lights. To highlight the feathery foliage of blue fescue or to light any white flowers, use blue mercury vapor lamps.

Bear's breech (*Acanthus mollis*)
Lily-of-the-Nile (*Agapanthus* species)
Elephant's ear (*Alocasia macrorrhiza*)
Heartleaf bergenia (*Bergenia cordifolia*)
Delphinium (*Delphinium elatum*)
Fortnight lily (*Dietes vegeta*)
Foxglove (*Digitalis purpurea*)
Globe thistle (*Echinops exaltatus*)
Japanese aralia (*Fatsia japonica*)
Blue fescue (*Festuca ovina* 'Glauca')
Plantain lily (*Hosta* species)
Red-hot poker (*Kniphofia uvaria*)
Lily turf (*Liriope* species)
Philodendron species
Yucca (*Yucca gloriosa* and other species)

10. Shrubs and Perennials with Attractive Texture
Some of these plants are shrubby, others prostrate and herbaceous, but all can be lit by low path or area lighting fixtures to bring out their texture. Light massed plantings of these species with grazing light directed across the surface.

Bearberry manzanita (*Arctostaphylos uva-ursi*)
St. Johnswort (*Hypericum calycinum*)
Juniper, low-growing species (*Juniperus horizontalis*, *J. procumbens*)
Mondo grass (*Ophiopogon japonicum*)
Japanese spurge (*Pachysandra terminalis*)
Stonecrop, low-growing species (*Sedum* species)

11. Plants with Translucent Leaves
The filmy leaves of these trees, shrubs and perennials can seem to glow when lit by fixtures at close range. For ferns and perennials, use very small uplights positioned at or near ground level. For trees, use uplights placed in the tree crotch, or downlights shining from nearby branches.

Southernwood, silver spreader (*Artemisia* species)
Mexican orange (*Choisya ternata*)
Many ferns (especially *Woodwardia*, *Athyrium*, and *Dicksonia* species, and others)
Empress tree (*Paulownia tomentosa*)
Golden bamboo (*Phyllostachys aurea*)
London plane tree (*Platanus* x *acerifolia*)

GLOSSARY

A-type lamp: A type of incandescent lamp; the common household light bulb.

Absorption: A measure of the amount of light striking an object that is absorbed rather than reflected. Surfaces that are black or dark-colored, have a dull finish, and are heavily textured absorb the most light. See also Reflectance.

Accent lighting: The use of small spotlights, decorative metal lamps, or other lighting fixtures designed to accent small garden areas.

Accent lighting

Ambient luminescence: Richard Kelly's term for shadowless, background illumination in an environment. An example is the fluorescent lighting in many offices, or fill lighting and soft area lighting in a garden.

Ampere: A measure of the flow of electrical current through a wire or other conductor.

Area lighting: The lighting of large garden areas for entertainment, general illumination, or games. Floodlights are frequently used for this purpose.

Axis: The main line of sight in a garden, which forms its basic structure.

Background lighting: Lighting of a wall, tall shrubs, or other vertical garden element as a background for other lighting.

Ballast: An electrical device that is used with fluorescent, mercury vapor, high-pressure sodium, and metal halide lamps to provide the power to start the lamp and to control the flow of electricity while it is operating. The ballast is usually built into the lighting fixture.

Box: As used in wiring, a metal or plastic box used to contain wire connections, switches, or receptacles.

Brightness: Also called illumination, this is the amount of light striking a surface or object, measured in footcandles.

Cable: Multiple wires arranged in a common covering of insulating plastic or other materials, used as an electrical conductor.

Circuit: The path of electrical current leading from a power source, through electrical switches, receptacles, fixtures, and appliances, then back to the source.

Circuit breaker: A mechanical device that opens to interrupt a circuit when the current in the circuit exceeds its safe maximum.

Conduit: A metal or plastic pipe used to enclose electrical wires and cables.

Contour lighting: Lighting of the "edges" between different plants and materials in the garden to emphasize landscape forms and structure.

Cross lighting: Illumination of an area or object from two or more points. Frequently used in downlighting, moonlighting, and the lighting of statuary, it creates softer shadows and more pleasing light than illumination from a single point.

Current: The flow of electricity through a conductor, such as a wire, measured in amperes.

Diffused lighting: Lighting employing diffusing fixtures. It is useful in avoiding glare, especially when fixtures must be located in the line of sight. See also Diffusion.

Diffusion: The scattering of the light emitted by a lamp, by means of a diffusing coating on the lamp, a lens or grill on the fixture, or other means. Diffused light is a softer light with less glare, and produces soft shadows.

Dimmer: A switch that dims the brightness of lighting fixtures by reducing the flow of current.

Dormancy: In plants, a period of rest or quiescence, usually initiated by cold weather.

Downlighting: The lighting of an object, area, or surface from above.

Energy efficiency: In lighting, energy efficiency is figured according to how much light (measured in lumens) is produced by a lamp for each watt of electricity supplied.

ER lamp: Ellipsoidal reflector lamp, a type of incandescent lamp. When using outdoors, install in a fixture with a protective lens.

Fill lighting: Dim lighting of a garden background and of areas adjacent to brighter spotlighting or area lighting, to provide contrast.

Fish tape: A flexible band of spring steel on a reel, used for pulling wire or cable through conduit.

Floodlight: A lamp that produces a bright, broad beam of light.

Fluorescent lamp: A lamp in which a coating on the inside of a glass tube is made to glow by an electrical current. Fluorescent lamps require special fixtures and ballasts, and give an even, glare-free light. See also the chart on page 6.

Focal glow: Richard Kelly's term for the bright light focused on particular elements in an environment. In the garden, spotlights and some types of accent lights produce focal glow.

Footcandle: The unit used to measure the brightness of the light striking a surface. One footcandle is the illumination falling on a surface of one foot square from a standard candle located one foot away.

Fuse: A safety device containing a band of metal that melts to interrupt a circuit when the current in the circuit exceeds a predetermined safe level.

GFI: See Ground fault circuit interrupter.

Glare: Distractingly bright light that interferes with our seeing what we need or want to see in an environment.

Grazing light: The positioning of a light source to bring out the texture or surface dimension of a wall, door, or other element.

Grazing light

Ground fault circuit interrupter (GFI): A safety device that interrupts a circuit in about $\frac{1}{40}$ of a second if it detects any leakage of current, to prevent danger of shock. Required by electrical codes on some parts of outdoor circuits. See also page 41.

Halogen lamp (Tungsten halogen lamp): See Quartz incandescent lamp.

High-intensity discharge lamp (HID lamp): A lamp that produces light when electricity excites specific gases within a pressurized bulb. This group includes mercury vapor, metal halide, and high-pressure sodium lamps. All high-intensity discharge lamps require special fixtures and ballasts.

High-pressure sodium lamp: A high-intensity discharge lamp that illuminates by radiation from sodium vapor. Its color rendering is distinctly yellow-orange. High-pressure sodium lamps require special fixtures and ballasts. See also the chart on page 6.

Illumination: See Brightness.

Incandescent lamp: A lamp that produces light when electricity heats a metal filament to incandescence. This is the standard household lamp, emitting a yellow-white light. See also the chart on page 6.

Junction box: An electrical box in which several wires are connected. It usually has a solid cover.

Kilowatt-hour (KWH): Unit of measure for electrical energy. One kilowatt-hour equals 1000 watts of electricity used for one hour, or 100 watts for 10 hours, and so on.

Lamp: The technical name for what we commonly call a light bulb. It is a tube, usually of glass, in which a filament, gas, or coating is excited by electricity to produce light.

Light bulb: See Lamp.

Light output: The amount of light emitted by a lamp, measured in lumens.

Light source: The combination of a lamp and fixture that illuminates an environment.

Lighting fixture: The housing for a lamp, usually containing a reflector and electrical wiring connected to a power source. It may also contain a lens to protect the lamp, and a ballast if it is intended for use with high-intensity discharge lamps.

Lighting system: The complete system of lamps, fixtures, wiring, switches, and a power source, that supplies light to an environment. May also include auxiliary equipment such as transformers, time clocks, and sensors.

Low-voltage lighting system: A type of lighting that operates on 12-volt current rather than the standard 120 volts. (A few systems use 24-volt power.) Power is supplied by a transformer, which itself is connected to 120-volt power.

Lumen: A unit measuring the amount of light emitted by a lamp. One lumen is the amount of light emitted by one standard candle. Lumens of emitted light are measured at the light source.

Mercury vapor lamp: A high-intensity discharge lamp that produces light by radiation from mercury vapor, when supplied with electricity. "Clear" (also called "blue") mercury vapor lamps are the main type used in garden lighting, and have a distinct blue-green color rendering. All mercury vapor lamps require special fixtures and ballasts.

Metal halide lamp: A high intensity discharge lamp that produces light by radiation from compounds of certain metals. The color rendering is greenish-white. Metal halide lamps require special fixtures and ballasts.

Mirror lighting: The use of a swimming pool or garden pool as a reflecting surface, by lighting nearby landscape elements while leaving the water dark.

Mirror lighting

Moonlighting: A type of lighting using a mild source of light, usually located high overhead, to suggest moonlight. Frequently used to light large specimen trees in the landscape, and to cast soft, shadowed light from the trees onto the garden below.

Overload: A situation in which the demand for current exceeds the capacity of the circuit or equipment. An overload usually causes the fuse or circuit breaker controlling the circuit to blow.

PAR lamp: Parabolic aluminized reflector lamp, a type of incandescent lamp made with Pyrex glass to withstand outdoor conditions.

Photocell: A device that responds electrically to a certain intensity of light. Photocells are used to automatically switch lighting systems off with daylight, on with darkness.

Play of brilliants: Richard Kelly's term for the stimulating, exciting accent light that brings an environment alive. Strings of mini-lights and some kinds of small accent lights produce this kind of light in a garden.

Quartz incandescent lamp: A type of incandescent lamp containing a tungsten filament within a pressurized fused-quartz bulb that is filled with a halogen gas. Quartz lamps emit a bright, yellow-white light.

R lamp: Reflector lamp, a type of incandescent lamp designed for indoor use. When used outdoors, install in a fixture protected by a lens.

Receptacle: A device that is installed in an electrical box to supply power to the electrical plug and cord of a fixture or appliance; an outlet.

Reflectance: A measure of the amount of light that strikes a surface and is reflected. Reflectance is highest from objects or surfaces that are shiny, smooth, and light-colored.

Security lighting: Lighting to protect a home from unwanted intruders. Security lighting is most effective when it eliminates shadows near house windows and doors, and lights the landscape floor evenly.

Shadowing: The projection of the shadow of a plant, frequently in exaggerated size, on a wall or other vertical surface.

Short circuit: An improper connection that interrupts an electrical circuit.

Silhouetting: The lighting of a vertical surface to silhouette the plants or objects in front of it.

Spotlight: A lamp producing a bright, highly focused beam of light.

Spotlighting: The use of spotlights and fixtures to illuminate specific garden elements and surfaces.

Standard-voltage lighting system: Lighting fixtures and lamps that operate on standard 120-volt house current.

Switch: A device that is used to connect or disconnect power from a circuit or fixture.

Transformer: A device that converts current of one voltage into current of another voltage. In garden lighting, transformers are usually used to convert 120-volt current into 12-volt current for low-voltage lighting systems.

Uplighting: The lighting of an object or surface from below.

Vista lighting: The use of subdued, controlled lighting to frame a view.

Volt: A measure of electrical pressure.

Voltage drop: A loss of electrical current on a circuit, due to overloaded or excessively long wires. Frequently indicated by the dimming of lamps at the ends of wire runs.

Watt: A unit of measurement of electric power. The unit by which electric companies meter power supplied.

Wire nut: Also called a solderless connector, this is a small plastic device that can be fastened over the bare, joined ends of several wires to protect and insulate the connection.

Index

U.S. Measure and Metric Conversion Chart

	Formulas for Exact Measures				Rounded Measures for Quick Reference		
	Symbol	**When you know:**	**Multiply by:**	**To find:**			
Mass (Weight)	oz	ounces	28.35	grams	1 oz		= 30 g
	lb	pounds	0.45	kilograms	4 oz		= 115 g
	g	grams	0.035	ounces	8 oz		= 225 g
	kg	kilograms	2.2	pounds	16 oz	= 1 lb	= 450 g
					32 oz	= 2 lb	= 900 g
					36 oz	= 2-1/4 lb	= 1000 g (1 kg)
Volume	tsp	teaspoons	5.0	milliliters	1/4 tsp	= 1/24 oz	= 1 ml
	tbsp	tablespoons	15.0	milliliters	1/2 tsp	= 1/12 oz	= 2 ml
	fl oz	fluid ounces	29.57	milliliters	1 tsp	= 1/6 oz	= 5 ml
	c	cups	0.24	liters	1 tbsp	= 1/2 oz	= 15 ml
	pt	pints	0.47	liters	1 c	= 8 oz	= 250 ml
	qt	quarts	0.95	liters	2 c (1 pt)	= 16 oz	= 500 ml
	gal	gallons	3.785	liters	4 c (1 qt)	= 32 oz	= 1 l
	ml	milliliters	0.034	fluid ounces	4 qt (1 gal)	= 128 oz	= 3-3/4 l
Length	in.	inches	2.54	centimeters	3/8 in.	= 1 cm	
	ft	feet	30.48	centimeters	1 in.	= 2.5 cm	
	yd	yards	0.9144	meters	2 in.	= 5 cm	
	mi	miles	1.609	kilometers			
	km	kilometers	0.621	miles	12 in. (1 ft)	= 30 cm	
	m	meters	1.094	yards	1 yd	= 90 cm	
	cm	centimeters	0.39	inches	100 ft	= 30 m	
					1 mi	= 1.6 km	
Temperature	°F	Fahrenheit	5/9 (after subtracting 32)	Celsius	32°F	= 0°C	
					68°F	= 20°C	
	°C	Celsius	9/5 (then add 32)	Fahrenheit	212°F	= 100°C	
Area	in.^{2}	square inches	6.452	square centimeters	1 in.^{2}	= 6.5 cm^2	
	ft^2	square feet	929.0	square centimeters	1 ft^2	= 930 cm^2	
	yd^2	square yards	8361.0	square centimeters	1 yd^2	= 8360 cm^2	
	a	acres	0.4047	hectares	1 a	= 4050 m^2	